W9-BNN-586

# The Marine Mammals of the Gulf of Mexico

NUMBER TWENTY-SIX:

*The W. L. Moody, Jr., Natural History Series*

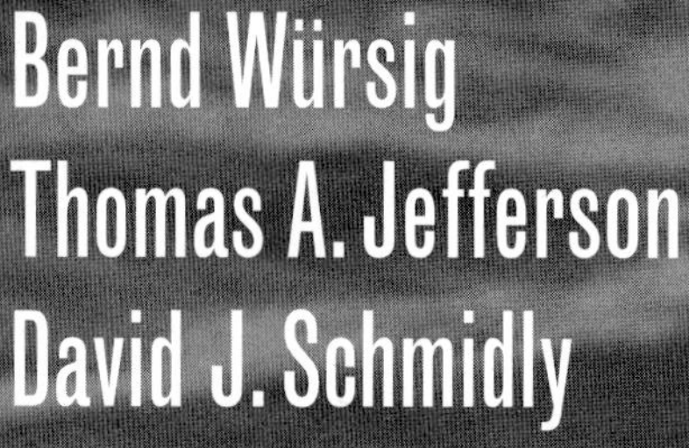
Bernd Würsig
Thomas A. Jefferson
David J. Schmidly

Cetacean Paintings by Larry Foster

TEXAS A&M UNIVERSITY PRESS
COLLEGE STATION

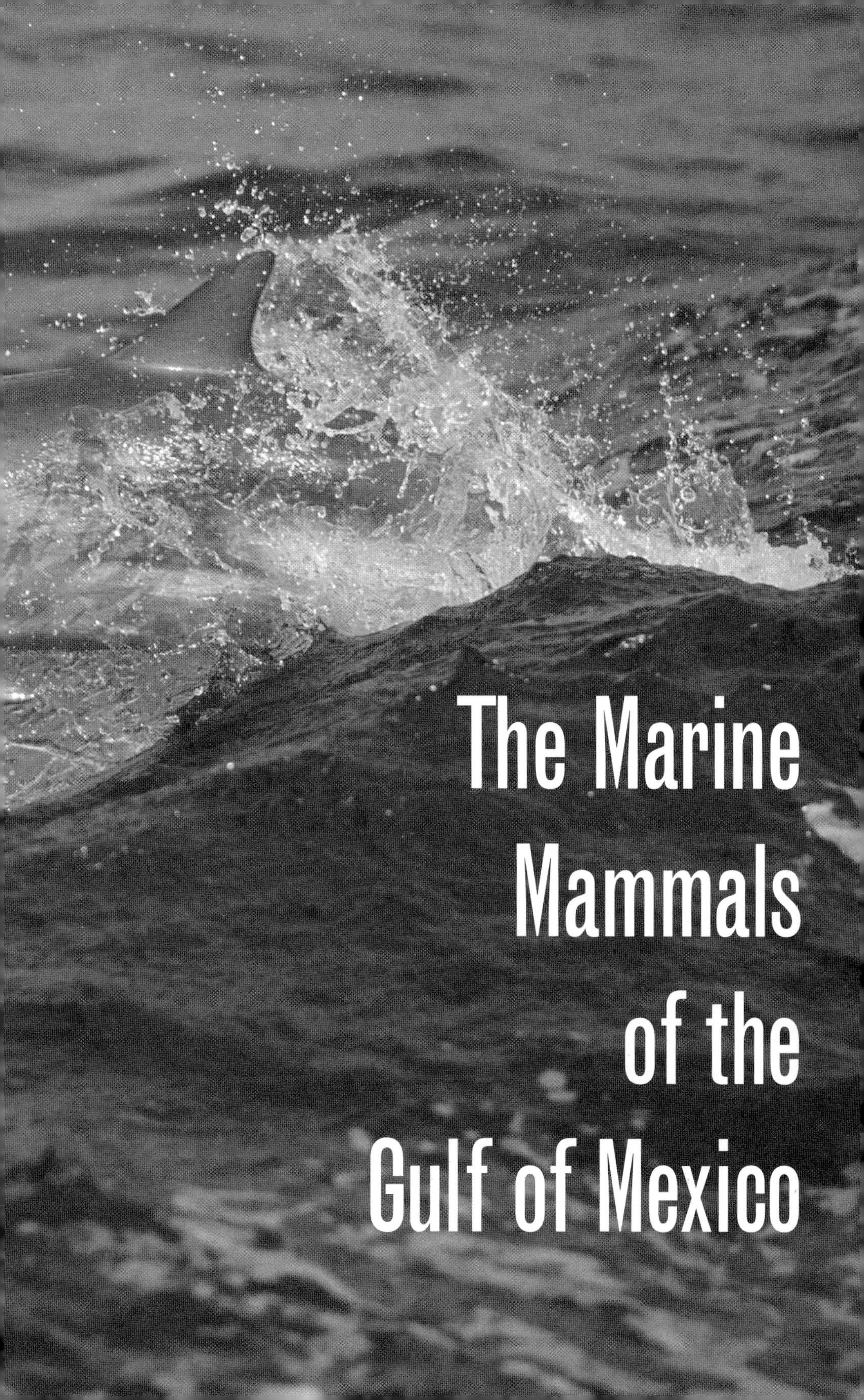

# The Marine Mammals of the Gulf of Mexico

Manufactured in the United States of America

First edition

*frontispiece: Stenella clymene,* Clymene dolphin.
Photograph by Keith D. Mullin

All cetacean paintings are by Larry Foster
except *West Indian Manatee* which
is by Jenny Markowitz.

The paper used in this book meets the minimum requirements of the American National Standard for Permanence of Paper for Printed Library Materials, Z39.48-1984. Binding materials have been chosen for durability.
∞

Library of Congress Cataloging-in-Publication Data

Würsig, Bernd G.
The marine mammals of the Gulf of Mexico / Bernd Würsig, Thomas A. Jefferson, David J. Schmidly ; cetacean paintings by Larry Foster. — 1st ed.
p. cm. — (The W.L. Moody, Jr., natural history series ; no. 26)
Includes bibliographical references.
ISBN 0-89096-909-4
1. Marine mammals—Mexico, Gulf of. I. Jefferson, Thomas A. II. Schmidly, David J., 1943– . III. Title. IV. Series.
QL713.2.W87 2000
599.5'177364—dc21 99-36385
CIP

This book on the marine mammals
of the Gulf of Mexico
is dedicated to the memory of
Dr. J. Stephen Leatherwood:
naturalist, scientist, conservationist,
colleague, and friend.

Dear Rebecca –
May this little book
always remind you of the
beauty of these "beasts," and of
our friendship. Love –
Bel
28 Mar. '00

# Contents

# Illustrations

## Photo gallery

## Plates

# Tables and Figures

## Tables



## Figures

# Prologue

When David Schmidly (DS) started this book more than ten years ago, knowledge about the marine mammals of the Gulf of Mexico came mainly from strandings of dead animals—individual carcasses lying on some often desolate shore. Smelly, rotten flesh was all that was left of an individual that only days before, or at most weeks, was an integral member of a close-knit society of a dolphin, whale, or sea cow species. From these remains alone, the untrained observer might find it difficult to imagine that these animals had lived sophisticated social lives, and that they experienced kinship and friendship relationships similar to those of all social mammals. If the carcass was that of a bottlenose dolphin, as was most often the case, it could be surmised that in life, it roamed through tight, grass-lined channels all along the shores of the Gulf of Mexico; if a manatee, it represented a marvelous fringing range of a tropical vegetarian sea cow; if the carcass was that of a dolphin or whale from deeper water, one knew only that the animals were out there, but absolutely nothing about their ranges in the deep waters of the Gulf of Mexico, their group sizes, social affiliations, underwater calls, reproductive patterns, or their problems.

We now know somewhat more than we knew then. DS established a comprehensive volunteer marine mammal stranding network along the shores of Texas in the 1970s, and began to compile data on strandings and record anecdotal sightings throughout the Gulf. From this, we learned that bottlenose dolphins are the most common cetaceans near shore, and we surmised (correctly) that sperm whales are year-round residents in the Gulf of Mexico, perhaps foregoing their usual migrations in these warm temperate to subtropical waters. It was not until dedicated surveys in the deep waters of the Gulf, sponsored mainly by the National Marine Fisheries Service in the 1980s and the Minerals Management Service in the 1990s, that we began to obtain an appreciation of habitat-use characteristics and the relative numbers for many of these inaccessible (via small boats from shore) creatures that were rarely documented in stranding records. Thus, we learned that Fraser's and Clymene dolphins—about which we knew very little from the stranding record—occur regularly in the Gulf. We also learned that common dolphins, now classified as two distinct species of the genus *Delphinus* found in and near tropical waters "worldwide," surprisingly are not common, and possibly absent from Gulf waters. Baleen whales are also less common than we had thought, and sperm whales even more common than we could have imagined.

We believe that we can now provide a credible account of some of the major species in the Gulf of Mexico. We, however, must present a set of caveats. First, our studies of the marine mammal fauna of the Gulf of Mexico are only beginning, and in ten or twenty years, some of the statements and suppositions we present may appear laughable. That is the nature of science. We accept our potential fallibility; the reader must accept that the science of counting and explaining marine mammals—which so often hide important parts of their lives from us beneath the waves—is lamentably inexact. Second, although we now have some information on the marine mammals of the northern or U.S. portion of the Gulf of Mexico, we have only scratched the surface in research of the southern, more tropical waters of the Gulf belonging to Mexico; we know even less about the deep waters adjacent to the western fringes of Cuba.

Therefore, *The Marine Mammals of the* Northern *Gulf of Mexico* may be a more appropriate title; however, we are brash enough to believe that we have sufficient data to represent the entire Gulf, realizing that our knowledge of the south is less complete. For example, we predict that we will find more of the Bryde's whale, a tropical baleen whale, in the south, find several more concentrations of sperm whales along the deepwater margins of the south–central Gulf, and gain better knowledge of the mysterious deepwater species of so-called beaked whales. We have recently begun a fruitful collaboration with Mexican marine mammal scientists that soon will lead us to a better understanding of southern waters. We will perhaps incorporate this knowledge into a future edition of this book.

A summary text such as this obviously relies on the help and wisdom of many people, and any attempt at acknowledgments is bound to be woefully inadequate. We thank Suse Shane and Betty Melcher, who helped DS put together the first compilation of strandings and sightings in the 1970s. Dr. Shane was also the first to describe the lives of bottlenose dolphins near Texas shores, and she has gone on to give us major insights into bottlenose dolphin and pilot whale social systems. In 1980, the stranding network was expanded and organized to include full-time personnel. The staffing included veterinarians and people trained in animal necropsy. We thank personnel of the Texas Marine Mammal Stranding Network, especially former head Raymond Tarpley and present head Graham Worthy, as well as coordinators Ann Bull and Lance Clark. Without these good folks, we could not continue to receive information on live and dead animals cast on shore.

Numerous others have also made important contributions. Dave Scarbrough labored on early drafts of species descriptions, which we have adapted. Larry Foster kindly supplied all of the cetacean drawings, and Jennifer Markowitz drew the manatee. Keith Mullin and the late Steve Leatherwood, both instrumental in the descriptions of marine mammals of the Gulf, added to our knowledge in innumerable ways and kindly reviewed portions of this manuscript. Randy Davis oversaw, and continues to oversee, one of the largest attempts ever at combining marine mammal sightings with detailed knowledge of oceanographic and environmental parameters, the "GulfCet" projects of the Gulf of Mexico. This work has been made possible by the U.S. Minerals Management Service (MMS), the National Biological Service (NBS), and the National Marine Fisheries Service (NMFS). We are especially thankful to Bob Avent and Dagmar Fertl of MMS, and to Keith Mullin, Larry Hansen, Gerry Scott, and the late Ben Blaylock of NMFS.

Our colleagues, Bill Evans, Jeff Norris, and Troy Sparks, taught us about sounds of marine mammals of the Gulf. Logan Respess ably guided a team of oceanography data gatherers, and Giulietta Fargion explained oceanographic parameters of the Gulf. Spencer Lynn, Bill Stevens, and David Weller helped Thomas Jefferson (TJ) and Bernd Würsig (BW) lead admirable teams of undergraduates and interns to sight marine mammals in often stomach-churning, rolling seas. Andy Schiro, Cheryl Shroeder, Spencer Lynn, and David Brandon saw to it that data were analyzed and presented in fair and readable fashion. Boats and planes and fine skippers and crews were also integrally involved. Dozens of students and helpers volunteered their time to recover strandings, make sightings of mammals near shore and on the high seas, and engage in the drudgery of entering seemingly endless rows and columns of numbers onto computer spreadsheets.

Joel Ortega-Ortiz helped fill in our scant knowledge of the southern Gulf. Dagmar Fertl provided invaluable updates on the literature of the Gulf. Karen St. John, Nicolle Dailey, and several fine interns in the laboratory of the Texas A&M University Marine Mammal Research Program helped put some of the final touches to the manuscript. Robert L. Brownell, Jr., reviewed select portions of the manuscript and provided advice on nomenclature, taxonomic presentation, and biology. Thank you.

This book is organized as a series of subject matter sections or chapters. The first chapter contains a general introduction to the marine mammals of the world; a global view is a prerequisite to understanding the more "local" Gulf of Mexico. The basic biology of the species is reviewed in this section, with special emphasis on the methods that different marine mammals use to dive and feed. The second chapter includes descriptions of some of the physical and biological characteristics of the Gulf; termed the "physiography and biology" of a region, they are crucial in any attempt to understand the marine mammal species that inhabit this area. The third chapter includes a list of all marine mammals known or suspected to occur in the Gulf, with keys to the identification of cetaceans, and accompanied by drawings and illustrations. Chapter four also includes a brief synopsis of past marine mammal studies in the Gulf. The fifth chapter presents the species descriptions for those marine mammals known or believed to occur in the Gulf, accompanied by photographs. This is the "meat" of the book, and what the reader should turn to in order to understand more about a particular species. The last chapter deals with anthropogenic problems of marine mammals throughout the world, with a discussion of environmental problems as we understand them in the Gulf of Mexico. We must first acknowledge the specific problems before we can begin to address potential solutions. Finally, we have included a list of suggested readings that is divided into general and specific sections. We have placed photographs of Gulf marine mammals throughout the text in the belief that no amount of words can substitute for a representation of the real animals in nature. We wish that we could animate them for you, and give them the beautiful breath of life they had when the photographer snapped the shutter. We were fortunate enough to obtain some of the best photos from some of the best marine mammal photographers who were ever in the Gulf; in many of these photos, the breath of life still seems to be there. We thank particularly Dagmar Fertl, Bob Pitman, Keith Mullin, Mike Newcomer, Carol Roden, and Jon Stern for their beautiful photos. Perhaps this book will convince you to visit the shore or even travel to the deep sea of the Gulf, in order to glimpse some of these marvelous creatures for yourself. If it does, our efforts will have been a success.



Bernd Würsig, Muritai, Kaikoura, New Zealand
Thomas Jefferson, Discovery Bay, Hong Kong
David Schmidly, Lubbock, Texas
MAY, 1999

# The Marine Mammals of the Gulf of Mexico

# Chapter 1

# Introduction to the Marine Mammals of the World

MARINE MAMMALS are those that carry out all or a substantial part of their foraging in marine, or in some cases, freshwater environments. The term marine mammal is purely descriptive and is not a taxonomic designation, encompassing mammals from three orders (Sirenia, Cetacea, and Carnivora) and at least five evolutionary lineages on land. This "hodgepodge" group includes mammals that compromise behavior and morphology between land and sea (polar bears, sea lions, fur seals, true seals, and walrus), and those emancipated from land by their ability to give birth in water (sea otters, sea cows, whales, dolphins, and porpoises). Certain species of even those oceangoing mammals, however, may come close to land to: graze on grasses lining the shore (sea cows); use protective shallow coves to rest (certain dolphins and newborn whales); feed on shallow-bottom invertebrates (otters, certain dolphins and porpoises, and gray whale); temporarily beach themselves and snap up fish that they had pushed onto land (bottlenose, hump-backed, and Amazon River dolphins); or grab sea lions or seals in the turbulent surf zone (killer whales).

All, however, are bound to the sea by the need to feed, and all have undergone remarkable adaptations to live at least part of their lives in a dense, heat-sapping, three-dimensional world. Many are excellent divers with formidable breath holding, thermoregulatory, pressure resistant, and sensory capabilities to cope with the rapid change in conditions from the water's surface to the cold and dark abyss below. There are about 120 of these remarkable mammals, and we will discuss in some detail the approximately one-fifth of this assemblage that have been found in the Gulf of Mexico. No sea otters, sea lions, fur seals, walrus, or true seals occur naturally in the Gulf, although they did millions of years ago. Very recently, the last remnant populations of the West Indian monk seal have died out, due certainly to human presence.

The objective of this introductory chapter is to give, as thoroughly as possible, a description of the lives of these remarkably diverse sea mammals. Constraints to brevity, however, permit only brief descriptions on general topics (taxonomy, evolution, morphology, behavior, and social systems) designed to provide the interested naturalist with a better overall appreciation of the marine mammals of the Gulf of Mexico and the rest of the world.

## Taxonomy, Evolution, and Basic Morphology

### ORDER SIRENIA

The oldest of the marine mammals are the sea cows of the order Sirenia. They evolved from early (proto) ungulates more than 70 million years ago (70 mya). Skull morphology and protein affinity studies of present-day manatees and the dugong indicate that the closest living relatives of sirenians are elephants (order Proboscidea) and hyraxes (order Hyracoidea). Although the exact sequence of events remains shrouded in mystery, it is not difficult to imagine that an early ungulate, perhaps not unlike in structure and behavior to modern hippopotamuses, grazed at first on the productive water's edge and then more and more in the water itself. Perhaps due to predation pressure on land, the most successful of these early marine mammals were the ones that minimized their time on shore, until finally they could mate, give birth, nurse, and carry out the entirety of their lives without leaving the water. This is not to be thought of as a quick evolutionary step; it is likely to have taken tens or hundreds of thousands of years to accomplish, and millions of years to perfect.

In the process, sirenians developed thick layers of insulating fat, or blubber, lost what had become useless hind limbs and most hair, and developed a powerful, large paddle-shaped tail. Their forelimbs became flattened and shortened, evolving into efficient structures that appear to be used mainly as steering paddles by the dugong and as food-manipulation devices by manatees. Being good descendants of ungulates, they remained herbivores. Sirenians are the only marine mammals that feed exclusively on plants, having retained the compartmentalized stomach that makes the grinding and processing of plant material possible. Their already dense and thick ungulate bones became even more dense (pachyostotic), an adaptation that allows them to browse along the bottom despite a prodigious, buoyant blubber mass.



Sirenians evolved in warm water, and their relatively slow metabolism and lethargic lifestyle of grazing in shallow water close to shore keeps present-day descendants from invading the cold waters of temperate climes. This, however, was not always the case. One presently extinct group of dugongs became very large and invaded the North Pacific, feeding not on tropical sea grasses but on cold-water kelp. The last species of this group, the Steller's sea cow, was discovered near an island off the far eastern coast of Russia by an expedition led by Vitus Bering. Adult Steller's sea cows were up to 9 m (29.5 ft.) long, slow, (therefore easily killed) and unfortunately, were said to have tasted "as the finest of veal." Subsequent explorers and fur hunters literally ate the animal to extinction, and twenty-eight years after its discovery in 1742, this magnificent cold-water offshoot of sirenians was extinct. All that remains are the manatees of western Africa, the Caribbean, and the Amazon, together with one species of dugong that occurs on the eastern coast of Africa and along the coasts of Saudi Arabia, Iran, Pakistan, India, southeast Asia, and northern Australia.

Where did sirenians evolve? To appreciate their evolution (and that of the next group, the cetaceans), it is important to realize that the world was not always as it is today, with seven continents and other large land masses. Due to the ongoing geologic process of sea floor spreading, more than 70 mya, the Atlantic was relatively narrow, with the Americas very close to the European and African masses to which they were formerly joined. The Pacific was larger than it is today. More importantly, however, the Mediterranean was also huge, consisting of a large ocean between Eurasia and Africa called the Tethys Sea. It was on the shores of the productive Tethys that many ungulates evolved and sirenians first invaded the sea. The dugongs followed the shores to the south and west as shorelines changed, and manatees populated the African shoreline. It is less clear how other manatees, and an offshoot of the dugongs that eventually led to the Steller's sea cow, crossed the Atlantic to establish themselves in the New World. They did so, however, when the Atlantic Ocean was not as wide as it is today. It is believed that ancestors of the Steller's sea cow invaded the Pacific through the Panama Seaway and became increasingly more cold-adapted as they moved north and grew in size. Alternatively, it is also possible that the Steller's sea cow is an offshoot of dugongs from southern Asia, but this hypothesis awaits further time-sequenced analyses. More has been deciphered (or hypothesized) regarding the details of movements by sirenians and the other marine mammals as related to shifts in current regimes, mass global temperature changes caused by glacial and interglacial periods, and other dynamic physical phenomena; however, the present detail is sufficient to convey a general appreciation of the continually changing status of species, their ranges, and extinctions. Indeed, for every marine mammal on earth today, many have evolved, flourished, and become extinct.



## ORDER CETACEA

The next oldest order is the Cetacea, which is comprised of two major suborders, the Mysticeti (baleen whales) and the Odontoceti (toothed whales). These two groups evolved from an even older group of ancient whales, the Archaeoceti.

Like sirenians, early whales also evolved from early ungulates, perhaps about 60 mya, when sirenians were already quite developed. Unlike the elephant–hyrax line that led to sirenians, however, archaeocetes evolved from a carnivorous group of ungulates called the mesonychid condylarths, which are now extinct. Cetaceans have to this day kept their carnivorous habits. Archaeoceti had heterodont (differentiated) teeth in their jaws for grabbing, tearing, and ripping prey. They did not have the chewing molars common to herbivorous ungulates, sirenians, and—for that matter—such omnivores as humans. Archaeocetes lost their external hind limbs, most body hair, and developed fore-flippers from forelegs and a flattened tail. There is evidence that early forms, around 45 to 60 mya, still had small hind legs. They were probably able to crawl onto land, perhaps to give birth—not unlike the obligate behavior of present-day sea lions and seals. Many later archaeocetes, however, were perfectly well adapted to a fully aquatic existence. They, like sirenians, also first expanded into the ocean in the Tethys Sea, and also may have been entering a new feeding niche while avoiding predators on land.

During this time, the oceans were relatively devoid of large flesh-eating creatures; the reptilian plesiosaurs and ichthyosaurs, which could have served both as competitors and direct predators of early archaeocetes, had died out several million years ago. Therefore, the oceans were theirs to invade as large predators; archaeocetes speciated and flourished for many millions of years. The fact that they were warm-blooded, unlike any other marine predators except for the much smaller marine birds, probably helped them outcompete the cold-blooded predatory fishes and coexist with sharks, which were limited to warmer habitats. The social structure and greater mental capabilities inherent to their mammalian ancestry also may explain why early whales (and present-day cetaceans) can successfully coexist with such fine predatory machines as large oceanic sharks.

Present-day baleen and toothed whales evolved from archaeocetes about 30 mya. Archaeocetes had paddle-shaped forelimbs, small or no hind limbs, probably large flattened tails, dorsal fins, little hair, thick blubber, and other morphological characteristics of modern whales, but also had heterodont (differentiated) teeth and a skull structure quite similar to their terrestrial ancestors. The mysticetes and odontocetes, however, are characterized by a peculiar evolution of the skull termed "telescoping." The telescoped skull consists of extremely elongated maxillaries and premaxillaries that have essentially shoved the nasal bones and attendant nostrils (blowhole[s]) toward the top of the head. At the same time, modern cetaceans no longer have heterodont dentition. Instead, the odontocetes show a homodont (all teeth alike) condition of rounded or pointed teeth. The mysticetes have lost all teeth (still seen in fetal baleen whales but resorbed by the time of birth), and developed an entirely new structure of ectodermal origin, the baleen, built of keratin, not unlike our fingernails and hair.

Early toothed and baleen whales had fundamentally different ways of making a living. Modern toothed whales still pursue prey in individual fashion, or snap up several fishes from a school. The baleen whales were and still are essentially batch or filter feeders, sieving huge stands of small invertebrate and vertebrate (schooling fish) prey into their giant nets of baleen. The nets are composed of rows of baleen plates hanging from the roof of the mouth to form a finely fringed network. Large nets are more efficient at capturing moving prey, but also require larger mouths and bodies for support. Several of the baleen whales are the largest creatures on earth and—as far as we know from the fossil record—the largest creatures ever to have inhabited earth.

The large whale sharks and basking sharks (the largest of all sharks) are also batch feeders and provide a fitting comparison to the mammalian filter feeders. While baleen whales "invented" a new feeding technique that provided them food resources (essentially without competition), what special advantage allowed toothed whales to diversify and flourish? The record is not totally clear, but it is very likely that the sophisticated use of sound by toothed whales developed very early in their evolutionary history. Present-day toothed whales echolocate by sending out short-duration, extremely high-frequency (usually) sounds from their heads, and receive the echoes of these sounds, bounced off objects ahead of them, with their lower jaws hinged close to the ears. The duration between sending and receiving sounds gives distance information to the toothed whale brain, and tonal differences in the sound between outgoing and incoming signals give information on shape and composition of the sound-reflecting object. This echolocation system is so sophisticated that odontocetes essentially "see" with it, and a large part of their brain is taken up by the need to process rapid-fire click trains used to search their environment. This capability allows them to scan the environment at night, in murky waters, and in unlit depths. They even may be able to scan the insides of many objects, such as fishes and other whales. This echolocation ability undoubtedly helped make modern odontocetes a successful group, but it probably also was present, although perhaps in less sophisticated fashion, in the early ancestral forms. Interestingly, the other group of mammals to invade a three-dimensional habitat, the bats of the order Chiroptera, also developed sophisticated echolocation to detect prey, predators, and other objects in their flight paths.

From about 30 to 60 mya, archaeocetes dominated the mammalian fauna in the seas. Most archaeocetes were small, on the order of several meters in length, but at least one, *Basilosaurus,* was up to about 20 m long. "Suddenly," baleen and toothed whales appeared in the fossil record about 30 mya, and although it is now clear that the two new forms derived from the ancient whales, the exact transition lines have not been found. Archaeocetes, perhaps outcompeted by the new suborders, declined and finally disappeared approximately 22 mya. By that time, both mysticetes and odontocetes were well established with representatives in all oceans.

Today, there are four taxonomic families of baleen whales and nine families of toothed whales. While the separate baleen whale families make their living in fundamentally different ways, no such clear-cut differences exist in the toothed whales.

*Suborder Mysticeti*

The baleen whales include the families Eschrichtiidae, Balaenidae, Neobalaenidae, and Balaenopteridae (Appendix). The only extant eschrichtiid is the Pacific gray whale. (An Atlantic variety was hunted to extinction by early whalers and disappeared in the 1700s.) The gray whale feeds mainly by sucking (apparently by tongue action) invertebrate prey out of bottom sediments. It then surfaces with the mouth partially open and squirts water and mud out through the baleen plates, leaving food caught on the fringes of baleen inside the mouth.

The balaenid whales, or right and bowhead whales, sieve food out of the water column by opening their mouths wide and advancing, like a plankton net pulled through clouds of invertebrates. Bowhead and right whales have the longest baleen plates of any whale, presenting a large surface area of mouth to the aggregated prey. The baleen fringes are also the finest of any whale, capable of trapping copepods much less than 1 cm in length. As a matter of fact, the slow-sieving nature of these whales makes it impossible for them to capture larger prey such as fishes and squid, because these prey are able to detect the slow approach of that huge gape and take evasive action.

The neobalaenid family consists of the pygmy right whale, a copepod and fish and squid feeding whale with a curved jaw like that of balaenids; it, however, has a dorsal fin and a slender body shape like that of another family, the Balaenopteridae. The pygmy right whale occurs only in higher latitudes of the southern hemisphere.

The balaenopterid whales are lunge feeders. They rapidly surge into aggregated invertebrates and even schooling fishes, taking enormous volumes of water and prey into their mouths at one gulp. To enhance their "bite size," their throats, equipped with longitudinal grooves running from the tip of the lower jaw to near the umbilicus, open up into immense rounded bags. This bag then contracts by muscular action and water is forced out of a partially opened mouth through the baleen plates; once again, food is caught on the internal fringes of baleen. The balaenopterids are commonly called "rorquals," a Norwegian term referring to the "tubes" (*ror*) of the lower throat, combined with the word for "whale" (*qual*).

To summarize, gray whales feed on bottom-dwelling, stationary prey. They are the only true "grazers." Right and bowhead whales sieve through the water column and take swimming prey. Rorquals actively lunge through often highly maneuverable larger prey, using the elements of speed and surprise to take their food. Their bodies betray their more active feeding style because they are sleek, relatively slim, and built for bursts of speed.

Most whales are distributed in both hemispheres, but are less prevalent in the tropics, a situation termed "anti-tropical distribution." The gray whale, however, is only found in the North Pacific, and the Bryde's (pronounced "bridees," with a short "i") whale prefers tropical waters. Gray whales do not occur in the Gulf of Mexico, since they were exterminated from the Atlantic several centuries ago. Northern right whales occur sporadically in the eastern part of the Gulf as accidentals that spend most of their time in the main part of the North Atlantic. Rorquals may be the most numerous, with blue, fin, Bryde's, and minke (pronounced "minkee") whales possibly found in the deepest parts of the Gulf, especially to the south.

As readily discerned from the foregoing, all cetaceans are collectively referred to as "whales." Some of these whales, however, are also termed dolphins, river dolphins, and porpoises, as described below.

*Suborder Odontoceti*

Toothed whales are recognized in nine taxonomic families (Appendix). The family Physeteridae includes the sperm whale, which is the largest of the toothed whales. The family Kogiidae is comprised of the so-called pygmy and dwarf sperm whales. The family Monodontidae includes the narwhal and beluga (or white) whales of the northern ice. The family Ziphiidae includes nineteen species of what are termed beaked whales, attributable to the generally thin beak abruptly leading to a bulbous forehead behind it.



The sperm whales, pygmy and dwarf sperm whales, and beaked whales feed mainly on squid, and interestingly, these three groups show reduced dentition; sperm whales have functional teeth only in the lower jaw, and most beaked whales have only one tooth, or tusk, in each of the two lower jaws. These tusks erupt from the gums of mature males and become secondary sexual characteristics that may be important in male-male competition for sexually receptive females. The beaked whale family is the least known of any of the marine mammals, and some beaked whales, apparently quite rare or elusive (or both) in the open seas, are known only from quite incomplete skeletons, a few brief glimpses, or a decaying carcass on the beach. Two "new" species of beaked whales have been described in the 1980s and 1990s, and this taxonomic family is one in which new discoveries yet may be made.

The largest odontocete family is the Delphinidae (dolphins and dolphinlike whales), with about 35 species worldwide. The largest of these "dolphins" is the killer whale, with adult males up to 9 m long. The smallest are several species of the southern hemisphere genus *Cephalorhynchus,* which are only 1.2–1.4 m in length. The delphinids comprise a diverse group, with species inhabiting shorelines, open ocean, and even some river systems.

The family Phocoenidae is comprised of six species of true porpoises, distinguished from dolphins by having generally blunter snouts, lower triangular dorsal fins, and spade-shaped instead of conical (dolphin) teeth. The true porpoises tend to prefer shallow, nearshore waters. (The Dall's porpoise of the North Pacific, however, can be found in oceanic waters, and the finless porpoise of Asia occurs both near shore and about 1600 km, or 1,000 miles, up the Yangtze River of China.)

Finally, there are five species of river dolphins that inhabit some of the major rivers of India-Pakistan, Bangladesh, China, and South America. These recently were split from one family to three: the Platanistidae (Indus and Ganges River dolphins), the Iniidae (Amazon River dolphin), and the Pontoporidae (Chinese river dolphin and Franciscana). Interestingly, there are no river dolphins in North America, Africa, Australasia, or Europe. The Franciscana of South America occurs in the Rio de la Plata, both along the shore in marine waters as far north as southern Brazil and several hundred kilometers south of the river along the coast of Argentina. It is therefore not truly of riverine habitats, but nonetheless closely related to the river dolphin groups.

The physeterid, kogiids, ziphiids, and delphinids have representatives in the Gulf of Mexico, but monodontids, phocoenids, and the three families of river dolphins are absent from the Gulf. Many of the toothed whales show an anti-tropical distribution pattern not unlike that of many mysticetes. Bottlenose dolphins, however, ubiquitous along the shores of the Gulf of Mexico, occur throughout the temperate zones of both hemispheres as well as in the tropics—both near and offshore. They are an almost "cosmopolitan" dolphin; their adaptability and wide range of habitat use has made them the best known, yet at the same time, one of the most frustratingly mysterious and complex of the cetaceans.

## ORDER CARNIVORA

### *Suborder Pinnipedia*

While the baleen and toothed whales were flourishing about 20 mya, sea lions, fur seals, and walrus-type creatures were making their first tentative forays into the sea. These members of the order Carnivora are commonly termed pinnipeds ("the feather-footed ones"), with the term pinnipeds variously referred to as an order, a suborder (the Pinnipedia) or as a non-taxonomic description of the group. The taxonomic designation "suborder of the order Carnivora" is used here, since it is the most widely accepted designation.

There are three taxonomic families of pinnipeds: the eared seals (Otariidae), comprised of sea lions and fur seals; the earless or true seals (Phocidae); and an intermediate group (Odobenidae) of which the walrus is the only living representative. All three families are clearly allied to the carnivores, but there is some debate as to what specific group(s) of carnivores gave rise to the pinnipeds. The prevailing wisdom, based on skull morphology, holds that otariids and odobenids evolved from bearlike (arctoid ursid) carnivores, whereas phocids came from mustelid stock. This view, however, has recently been challenged. The early fossil record of phocids (ca. 15–22 mya) is scarce, and phocids could just as easily have arisen from the bearlike stock of the otariids. At the same time, studies of proteins and genetic makeup indicate that pinnipeds as a group are very closely related, and the idea of monophyletic (one ancestry) origin of all pinnipeds is becoming more widespread. The final outcome of this phylogenetic argument, however, has not been resolved.

Fundamental morphological differences exist between the phocids and otariids. On one hand, phocids cannot fold their hind flippers under their bodies for locomotion on land. Instead, they shift the weight of their bodies to move in caterpillar-like fashion (often termed "gallumphing") on rocks, sand, or ice. Some species also are able to effect a sinusoidal type movement not unlike some snakes. Otariids, on the other hand, fold their hind flippers under their bodies and walk ("waddle" is perhaps a better term) by alternate movements of hind and front flippers. In water, phocids use their long "feathery" hind limbs to propel themselves, while short, stubby fore-flippers (not unlike flippers of cetaceans and the dugong) help steer the animals. Otariids, in contrast, use their large, flat forelimbs to propel themselves underwater (similar to penguins and sea turtles), and let the hind limbs stream behind for steering (but not propulsion). Walrus (Odobenidae) are in between these two groups, propelling themselves underwater by both front and hind appendages. All pinnipeds seem somewhat ungainly on land, but in water, where speed and dexterity are important to catch prey and avoid predators, they are graceful masters of their environment. Pinnipeds have compromised body morphology between land and sea, but have biased their efforts (in evolutionary terms) to life in the sea.

The other fundamental difference between phocids and otariids is the absence of external earflaps (or pinnae) in the former (hence the name "earless" seals) and their presence in the latter (the "eared" or true seals). The phocids and walrus rely mainly on body fat for insulation, whereas otariids rely both on body fat and fur. One group of otariids, the fur seals, which have a highly dense pelage, rely almost exclusively on trapped air in their fur to stay warm. Reliance on fur to keep warm limits these otariids to temperate and near-arctic (but not true arctic) conditions. It also restricts them to relatively shallow dives, because dives decrease the air layer in fur and cause fur insulation to become useless in cold, deep water. The phocids, on the other hand, have successfully invaded the high arctic and antarctic regimes of ice and snow. For example, Weddell seals of Antarctica are perfectly at home in -50°C air temperatures and water at just below 0°C.



Most evolution and differentiation of eared seals occurred in the North Pacific, and they radiated to the South Pacific, South Atlantic (from around the tip of South America), and subantarctic islands. They never invaded the North Atlantic, however, and sea lions and fur seals are thus absent from the Gulf of Mexico. The true seals differentiated in the North Atlantic and invaded the Pacific and the southern hemisphere by way of the Panama Seaway between North and South America. This radiation occurred during one of the seaway's open periods, in a manner similar to the much earlier movement of sea cows from the Atlantic to the Pacific. Phocids now occur in the North Atlantic, several northern lakes of Eurasia, the North and South Pacific, and Antarctica. Small relict forms persist in the tropics of Hawaii, the Mediterranean Sea, and until recently, occurred in the Caribbean. Because pinnipeds need to breed on land or ice, they are particularly vulnerable to adverse human influence, and it is becoming a major challenge to continue protecting these coastal members of the marine mammal fauna.

## OTHER MARINE CARNIVORES

Besides the pinnipeds, there are two other forms of marine mammals that more recently evolved from terrestrial carnivore stock. The sea and marine otters of the North and South Pacific shores of the Americas, respectively, are essentially very large river otters that do all or almost all of their feeding in the sea. The sea otters of the northern hemisphere can give birth in water, and let their young ride on the female's back and belly for the first several months of life. They evolved from river otters no more than about 2 mya.

The second of the non-pinniped types is both a fine marine and terrestrial mammal—the polar bear. Polar bears have been tracked (via a satellite transmitter on a collar placed around the bear's neck) as far inland as about 1000 km (620 mi.), where they feed on river fishes, foxes, caribou, and other land game. Polar bears also have been seen as far as 100 km (62 mi.) from the nearest land or ice, placidly swimming along with their powerful front and hind limbs and feeding on seals and arctic char and other fishes. Usually, however, polar bears are found on sea ice, where they can surprise seals resting near their breathing holes. They are the largest of the living carnivores, and have evolved to admirably fill a most challenging set of niches in a very hostile environment.

### *Diving*

All marine mammals dive, but not all are equal at this task. Sea cows, sea otters, gray whales, and walrus do much feeding on the ocean bottom; however, they generally are not divers, preferring to stay in water only tens, not hundreds, of meters deep. It has long been known that sperm whales are particularly deep divers, capable of foraging on oceanic squid more than 1 km (0.6 mi.) below the surface, and staying underwater for one hour or more. After a long dive, sperm whales need to stay at the surface for 10 to 15 minutes, breathing repeatedly in an attempt to rebuild strongly depleted oxygen stores. They may even build up a so-called oxygen debt because muscles shift to anaerobic metabolism during the prolonged dive. Other deep (and longtime) divers include many of the beaked and pilot whales, and others that habitually feed on deepwater fishes and squid. Interestingly, the baleen whales are probably not exceptionally deep divers, and neither are most of the small, pelagic delphinid cetaceans. Although detailed dive profiles have been obtained only for a few species, it is known that the remarkable bottlenose dolphin can reach depths of about 0.5 km (0.3 mi.) during human-trained dives.

The real interesting diver, and possibly the champion among the pinnipeds, turns out to be the largest of the true seals—the elephant seal. Females dive more than 800 m (2624 ft.), and the much larger males can dive more than 1200 m (3936 ft.). These dives are exceptional, but dives of around 0.5 km (640 ft.) can be made repeatedly without pause (with dive times of ca. 15 minutes and surface times of ca. 3 minutes) for hours, days, and weeks. It has been hypothesized that these dives serve not only to obtain food, but are for resting underwater and avoiding the more shallow-diving shark predators. At any rate, the feats of diving that elephant seals exhibit apparently do not place them into anaerobic metabolism (or else they would need to rest for longer times at the surface while breathing to replenish depleted oxygen stores); the complex adaptations that make these dives possible, however, have not yet been fully explained.

Dolphins and whales dive with full lungs, whereas pinnipeds generally exhale before a dive or as they begin to submerge. In all marine mammal divers, it is likely that little air exchange from lungs to body tissues actually takes place deeper than about 60 m (about 200 ft.). If it did, nitrogen would be forced into solution under pressure and likely bubble out of the blood upon ascent, creating a dangerous condition known as the bends. Diving marine mammals appear to exchange most oxygen at or near the surface, and then store the oxygen in solution; this is facilitated by a high concentration of hemoglobin in the blood (so high that seal and dolphin blood appears almost black) and an extremely high oxygen carrying capability of the muscles. Other adaptations maintain respiratory exchange at a minimum during deep dives—the alveoli of the lungs probably collapse under pressure, and the lungs collapse due to a flexible rib cage and an obliquely placed diaphragm. As air volume decreases during a dive, blood-absorbing bundles of arterioles and venules (the *retia mirabilia,* or "wonderful net") fill in the spaces, especially in the head and neck region, to prevent tissue compression. An extensive bundle of blood vessels centered around the spinal column of many cetaceans may also trap gas bubbles and release them more slowly upon ascent than in other swimming vertebrates.

### *Sound Production and Senses*

Underwater sight is often restricted by turbidity and depth. Sound, however, transmits quickly (4.5 times faster than in air) and with little attenuation (or loss of loudness) in an environment that is 800 times more dense than air. Thus in general, marine mammals would be expected to be sonic creatures par excellence, and this certainly has been verified. Sea lions communicate with barklike sounds underwater and seals emit various grunts, clicks, and moans. Baleen whale sounds include the somewhat monotonous clicklike sounds of gray whales, the long, low moans (infrasonic, or below human hearing frequencies) of blue and fin whales, and the complex, repeated series of notes (properly called songs) of humpback and bowhead whales. Much of the underwater sounds of whales and seals appear to be associated with mate acquisition. Gray whale clicks possibly serve not only for communication but as a primitive sort of echolocation system. The moans of blue and fin whales can travel tens (and even hundreds) of kilometers, and probably keep these whales in touch with each other over long ranges, perhaps even serving as long-distance echolocation systems. The songs of male humpbacks are part of their sexual display on breeding grounds.



The odontocete cetaceans take production and receiving of sounds to an even finer behavioral and anatomic art. Most produce two distinct sound types—pulsed clicks and frequency-modulated whistles. Both whistles and click trains are used for communication, but certain high-frequency click sounds are also used for echolocation. Echolocation allows these animals to be highly efficient predators, even in waters in which eyesight is severely limited; the Ganges and Indus River dolphins, having evolved in a highly turbid, sediment-laden environment, demonstrate dramatically reduced eyesight. In all fairness to the sense of sight, however, most other marine mammals have very keen eyesight (the dolphins in air as well as underwater), and some antarctic seals have extremely large light-collecting eyes for vision under ice in the six-month-long night of winter. On land, pinnipeds in general do not have very acute vision, relying instead on barking sounds (e.g., otariids and some phocids) for communication, and smell between mother and calf in breeding colonies. Manatees and dugongs produce chirps, whistles, and squawks that allow mothers and calves to stay together; they also may function as other social bonding and mating signals.

Pinnipeds produce underwater sounds by recycling air between their vocal chords and larynx. Cetaceans, however, do not have vocal chords, although it is believed that vibrations of the laryngeal region are responsible for sound production in mysticete whales. It is clear, however, that most sounds of odontocetes are produced not at the larynx, but instead by shunting air between sets of nasal sacs in the forehead; in the process, the tissues of a mass of muscle (the nasal plug) are vibrated. The exact means of producing click and whistle sounds is unknown, but many dolphins apparently are able to produce both clicks and whistles simultaneously, suggesting another production mechanism in the forehead. The entire process probably evolved from the sibilant nasal sounds made by ungulates on land, transposed during evolution to an internal shunting of air due to the necessity of keeping the nares closed underwater. Sounds are received by the sides of the head, with the air-filled passage of the outer ear having degenerated in the pinnipeds and cetaceans. In toothed whales, much sound energy is also picked up along the sides of the lower jaws, leading to an efficient separation of right and left hearing; this thereby allows for highly discriminatory directional hearing, especially of echolocation-type sounds.

### *Social Organization and Breeding Systems*

None of the marine mammals appear to be truly monogamous (i.e., mate for life). Polar bear males and females stay together for only the brief period of estrus, as do many seals living on ice. Sea otter males set up a maritory (a marine territory) near shore, and actively exclude other males while attempting to keep females in the maritory from straying. They thus show a type of social system termed polygyny, with one male having access to several females but not vice-versa. This is similar to the "harem-like" structure of many polygynous seals and sea lions in which males fight for status and actively defend a shore-lining territory containing aggregated females; males even may sequester females in an area. In elephant seals, only some adult males ever make it to "alpha bull" status, which accords the exclusive right to mate with females; most males go through life attempting, but never arriving, at such status. The system, however, is not as rigid as it may seem, because subdominant males "sneak" copulations from peripheral females or those entering (or leaving) the water; such furtive attempts may be important as alternative mating strategies. This echoes a theme found throughout the animal kingdom: when a mating system excludes many for the benefit of a few, alternative strategies weaken the overall rigidity of the dominant system. One general rule prevails—in highly polygynous pinniped species, males are much larger than females and generally have other secondary sexual characteristics. This sexual dimorphism is found throughout polygynous mammal systems.

Sirenian social systems and mating strategies are not well known. Dugongs often occur in large herds of tens to hundreds of animals, whereas manatees often are found singly or in smaller groups. Dugongs may have one male tending to an estrous female for several days; this male is presumed to be the exclusive mate during that time. This is far from clear, however, and may not pose a hard-and-fast rule. Manatees (at least the West Indian manatee) may have a dozen or more males surrounding an estrous female, a mating aggregation that superficially appears similar to sexual aggregations of humpback, gray, right, and bowhead whales (described below).



Polar bears, sea otters, and most pinnipeds are solitary or aggregate casually, except during mating. Whales in particular often travel alone while on the feeding grounds. All marine mammals except for polar bears and sea otters have only one young at a time. While twins have been documented rarely in marine mammals, it is apparently too energetically draining for sirenians, pinnipeds, and cetaceans to successfully care for more than one young at a time.

It was believed that baleen whales were generally monogamous, but this is clearly not the case for at least humpback, bowhead, right, and gray whales. These whales are often seen on mating grounds in active groups of rolling, touching, pushing animals, usually with several males and one or two females. For bowhead, right, and gray whales, we have witnessed sequential mating of one female with several males, and have seen the same female in different groups of mating aggregations within hours or days. This situation strongly indicates a multi-male, multi-female system termed polygandry (also polygynandry), denoting a mixture of polygyny (one male and several females) and polyandry (one female and several males). It should not be termed promiscuous, since promiscuity implies the absence of mate choice and it seems likely that mate choice is an important part of these mating aggregations. There are alternatives to the active mating groups, and humpback song (also bowhead whale song) probably helps males sort out dominance relationships. It is not known how any balaenopterid whales mate. It also is not known to what extent groups of baleen whales stay together on the breeding and feeding grounds. Most indications, however, are that individuals meet and aggregate casually and for the short term, with long-term social bonds (beyond the mother-calf bond of ca. 1 year) being relatively rare. Nevertheless, certain recognizable adult humpback whales have been seen feeding together in Alaska during subsequent years, suggesting more long-term social bonding than previously thought.

Toothed whales, being a large and diverse group, present special problems in attempts to encapsulate mating–social strategies. The real problem is that very little information is available for most species. Sperm whales have a matriarchal society with females staying in lifelong social bands; adult males travel with these bands only during female estrus. This strategy of polygyny appears very similar to that of elephants. In sperm whales and others such as pilot and beaked whales, matriarchal society may be linked closely to their need of diving deep for squid; if related females stay together, they can care for calves at the surface while the mother feeds below.

Killer whale societies are also stable, and certain adult male killer whales travel with females and young for life, or at least prolonged periods of time. Similar situations may exist for certain beaked whales, pilot whales, and others. A general theme of polygyny appears closely tied to different physical attributes between males and females of a species, or sexual dimorphism. Thus, killer and sperm whale males are much larger than their female counterparts, and beaked whale and narwhal males have tusks as adults. Some species of smaller odontocetes also show sexual dimorphism, but most appear remarkably similar (at least to our eyes).

The sexual monomorphism of the often large groups of pelagic delphinid cetaceans is indicative of a polygandrous system, and observations of females and males mating with various partners appear to support this. To what extent these partners are kept for long periods of time, or whether sexual affiliations are more widespread throughout a large school, is not known. Only paternity and maternity studies based on extensive genetic testing of the patterns of relatedness may help answer this enigma. Bottlenose, Hawaiian spinner, Indo-Pacific hump-backed, and dusky dolphins show a certain group openness, with many individuals changing affiliations on an almost daily basis. This suggests that the stable mating unit is found mainly at the overall population level; however, this assertion is far from clear.

*Movements and Migrations*



Many mammals move to different habitats, often within a confined home range, for the various activities of resting, feeding, and mating. Marine mammals are no exception, and the pinnipeds, as we have seen, radically change environments between feeding in the sea and breeding on land or ice. Some marine mammals, such as white whales and many sea lions, undergo extensive annual migrations to avoid extreme water temperatures, or to follow productive currents or seasonal migrations of preferred prey. Some, such as Hawaiian spinner and dusky dolphins, show daily movement patterns that alternately put them in touch with prey and nearshore resting areas. Some inhabitants of temperate zones (e.g., several bottlenose dolphin populations of the eastern U.S. coast) demonstrate what could be called a partial migration, whereby some members move to different areas on a seasonal basis and others remain squatters in a confined nearshore area for life. The determinants of partial migration are not known, although they may involve various, partially overlapping populations having different preferred prey. There are also differences in movements based on age and sex; for example, juvenile male bottlenose dolphins range farther than adults of either sex, and mature male sperm whales and elephant seals range into much higher latitudes than females and younger males. In polygynous societies, males generally travel greater distances in search of food (and often, mates) than females.

The real long-distance migrators, however, are some of the great whales, including the sperm and most baleen whales. Gray whales cover the longest distance between feeding and mating grounds of any mammal, from the frigid productive waters of the Bering and Chukchi Seas to their balmy winter habitats off the coast of Mexico. Most baleen whales give birth in the warmer waters of low latitudes during winter and feed in the more productive colder waters near the poles during summer. It has been surmised that they must leave the cold waters to give birth; however, because of the large size of most newborn whales (gray whale: 4 m long at birth; blue whale: ca. 7 m) and their concomitant greater volume-to-surface area ratio compared to many smaller cetaceans that give birth in cold waters, that explanation is not likely to be the only one. Whatever the reason(s) for migration, large whales are able to leave the productive high latitude waters because they have sufficient body fat reserves to survive through prolonged fasting periods.



How do humpback whales find the Hawaiian Islands in a huge ocean, gray whales find their breeding lagoons, and right whales meet at certain areas for food and mates? Nearshore migrators such as gray, bowhead, and right whales may simply follow the coastline or use the bottom topography as a geographic clue of sorts. Whales, however, also travel on the high seas without reference to coasts or bottom, and how they find their way is not really known. They may follow currents and particular water tastes. The position of the sun at particular times of the day, seasons determined by a built-in biological clock, the earth's magnetic cues, or other yet unidentified cues may serve as guides. The earth's magnetic map is a particularly likely candidate for facilitating orientation (the ability to tell direction) and navigation (the more complicated ability to resolve location on a map and move from one area to another). Many animals (from bees to pigeons) are known to use magnetic cues, and there is some evidence that fin whales, for example, may follow lines of magnetic minima in the world's oceans. The large baleen whales also produce long moans of low frequency; it has recently been suggested that their moans may bounce off bottom and distant land topographies, thereby serving as a long-distance echolocation system. When migrations become better understood, a multiplicity of cues—ranging from memory of specific areas to sampling of subtle environmental parameters—are likely to emerge as controlling factors.

### *Strandings of Cetaceans*

From time immemorial, humans near shore have witnessed whales and dolphins cast on land to die. When the animals come to shore dead, often singly, this can be explained as the currents and winds simply having drifted them to land; however, when they strand on shore alive, often as large groups whose members strand repeatedly if they are helped back to the sea, it becomes apparent that there are more profound large-scale physical oceanographic or biological reasons for the phenomenon. Many explanations have been proposed, but none have proven to be completely satisfactory. As in the search for migratory cues, the answers probably lie in a multiplicity of factors.



Live mass strandings are generally characteristic of animals that are infrequent visitors to shallow coastlines, such as sperm, pilot, and false killer whales. These species are also highly gregarious animals, staying together as a cohesive unit in life; their group integrity simply may dictate that if some strand for a particular reason, others follow. Especially severe parasite infections have at times been noted in mass stranders, often in the middle and inner ears; it has been hypothesized that they might disturb hearing and equilibrium. Stranders therefore may have lost the ability to echolocate or sense the shore; perhaps their debilitated sensory system caused them to misread possible magnetic cues, and they lost their way. Not all mass strandings, however, involve massively infected animals or occur near areas of magnetic anomalies. At times, mass strandings may be attributable simply to the animals becoming lost close to shore, and getting caught by developing sandbanks or mudbanks as the tide recedes. At this writing, it is not known why mass strandings occur, and it remains a mysterious and complex subject in need of further study.

***Perspectives about the Present and the Future***

During the past thirty years, knowledge regarding the marine mammals of the world (and thus the Gulf of Mexico) has advanced substantially. Almost every day brings another revelation, hypothesis, or discovery about these denizens of the sea. This book describes the present knowledge about the Gulf, which is so much better than the "educated" guesses of the past. What exactly are the really exciting findings of the present? There are probably as many answers to this as there are researchers studying marine mammals, but we do have several thoughts.

Marine mammals are creatures marvelously adapted to a quite foreign environment, but physiologically and behaviorally, they fit well within the general scheme of the "typical" mammal. Physiologically, it was believed that pinnipeds and cetaceans had very high metabolic rates, necessary for living in a heat-sapping environment; however, this assessment came mainly from measuring the heart rates, breathing rates, and oxygen/carbon dioxide exchange volumes of captive animals that were often strapped onto a laboratory table (pinnipeds) or beached at the bottom of a drained pool (dolphins) while scientists probed, poked, and measured. Metabolic rates would be expected to be high under these stressful conditions. Additionally, because of being smaller and easier to handle, younger animals having inherently higher metabolic rates were tested; as well they were often what physiologists call "absorptive," meaning they had recently fed and consequently exhibited elevated metabolism due to digestion.

When these factors are standardized in the lab and compared to measures of free-swimming animals, it becomes evident that there is nothing remarkable about the general physiology of marine mammals (at least on the surface). This finding is somewhat remarkable in itself; a 30-g mouse that breathes 100 times per minute and lives for 2 years experiences roughly the same number of breaths per lifetime (ca. 100 million) as a 60-ton blue whale that breathes 2–3 times per minute and lives for 80 years. Similarly, the ratio of breaths to heartbeats is about 1:4.5 (independent of size), and both animals experience the same "physiological time." (Bats and many other small mammals increase their life spans tremendously by periodically lowering the basic metabolic functions of heart and blood rates during torpor or hibernation.)

Behaviorally, marine mammals also share many traits with terrestrial ones. While some researchers and many writers have argued for exceptionally high "intelligence" in whales and (especially) dolphins, and concomitant sophisticated language capabilities, this is not likely to be true. Cetaceans at least are bright social mammals with varied and rapidly adaptive behavioral repertoires. Their behaviors in captivity and nature, however, indicate that they are not at all "intellectual super beings." They all have large brain/body size and weight ratios, although some are apparently "brighter," by morphological and behavioral measures, than others. For example, the river dolphins, porpoises, and several of the smaller delphinids might well be thought of as less behaviorally flexible than bottlenose dolphins, pilot whales, and killer whales. Perhaps a comparison to the world of primates is fitting, with some dolphins and porpoises more like tree lemurs and howler monkeys, and others more like the chimpanzees and other great apes.

One of these "brightest" of marine mammals at the far end of the spectrum is the rough-toothed dolphin, *Steno bredanensis.* By all measures (brain size, brain complexity, and behavioral), this bottlenose-sized dolphin is indeed the most intelligent creature in the ocean. Pinnipeds show their terrestrial ancestry well, and much of the behavior and communication of fur seals, for example, can be compared in broad strokes to some carnivores on land.



The intention here is not to make exact comparisons between certain species in the sea and on land. Although there are large variations in behavioral capabilities and repertoires, strong similarities exist between marine and terrestrial mammals. For example, the highly polygynous and social matriarchal structures of sperm whales and elephants are now reasonably well known. Both are large, long-lived, slowly reproducing (ecologists call them "K-selected") species that need to move over vast areas in search of food and mates. This combination of ecological and morphological characteristics may have helped define the convergence of lifestyles and mating strategies, but the exact determining mechanisms (on land or in the sea) are not yet clear.

Although marine mammals show strong similarities to their terrestrial counterparts, they also demonstrate the great flexibility inherent in species that can live in widely divergent ecosystems. Various dusky dolphin populations of the southern hemisphere at first glance appear to be "totally similar" in their haunts off Patagonia, Argentina, and South Island of New Zealand on the other side of the world. Indeed, their morphology and their basic patterns of surface-leaping and individual interactions are almost the same. In Argentina, however, dusky dolphins cooperatively hunt schooling fishes in shallow water, and in New Zealand, they feed on deepwater lantern fishes. In Argentina, they search for fish schools in small groups spread over a wide area and later coalesce to help each other (a "fission-fusion society"); whereas in New Zealand, they stay together or close to the large school of dolphins, always in a protective and constantly communicating unit. Subsequently, social bouts, family units, and mating strategies are different. Dusky dolphins in these two worlds are just as different in their ways as a human large-city dweller in a high-rise apartment close to hundreds of others, and a farming family separated by many miles from the next farmhouse. In the Gulf of Mexico, bottlenose dolphins of the offshore environment may have very different strategies of living from those that make their home among the extensive shallow bays and estuaries inshore. There are hints of different society structure (e.g., larger group sizes in more open waters) in this species, but no firm conclusions at this time.

The study of marine mammals began largely with the anatomy of dead animals, shifted to evaluating the physiology and behavior of live animals kept in captivity, and then moved to assessing the lives of these animals in nature. It is now possible to: use satellite telemetry and sophisticated underwater listening stations to track creatures all over the oceans; biopsy animals and tissues to determine population isolation, genetic relationships within a school, and pollution loads in body tissues; and use a wealth of other techniques besides the traditional ones of cutting open dead animals or only verbally describing them in life. New techniques have not made the original ones obsolete; instead, they add to the gathering of knowledge, often in a synergistic fashion. For example, it is common knowledge that the milk of elephant seals is incredibly fat-rich, which was viewed as a valuable trait in preparing a helpless newborn to survive in cold and dangerous waters for two or three months after birth. From studies in nature, linking physiological traits, behavioral associations, and social relationships, it is now known that those mothers which were younger, smaller, or had lower fat content in their milk, produced offspring that either did not survive as well or that themselves produced less viable offspring after they became mature. The next step is to determine how "healthier" fathers affect generations of offspring. It is suspected that they do, as has been demonstrated recently in several bird species; however, diligent work with genetics (at this time the sole truly reliable technique for determining paternity) is needed to provide more information about long-term reproductive success, as well as details about the myriad of sexual strategies employed by these mammals in the first place.

The amounts and types of discoveries on marine mammals are increasing almost exponentially, and every researcher will have pet examples of the most exciting new knowledge. A particularly interesting example involves a recent finding on the ocean bottom of the deep sea. It was discovered that large whales are not totally decomposed by bacteria and detritivores before they reach the sea bottom, or soon thereafter, as was once widely believed. Instead, they come to rest, often thousands of meters below their more surface-bound homes in life, and there they are slowly digested in the cold and light-sparse depths. While decomposing, perhaps for dozens of years, they support an entire community of animals that are nourished in large part by the same kinds (and many of the same species) of sulfur-reducing bacteria that live on deep-sea hydrothermal vents. The whale carcasses therefore may be serving as oases of life in a veritable desert, in which creatures relying on chemosynthetic-based life (i.e., the bacteria that feed on the fat-rich bodies of the whales) are able to colonize extremely dispersed hydrothermal vents. In this scheme, the whales are at least partly responsible for increased biodiversity in deep-sea basins, and large-whale evolution these past fifty million years or so may have been pivotal in such colonizations. To us, this intriguing possibility resonates loudly, especially since whales have been deemed to be unimportant in ocean ecology due to the relative low values of their ecological mass in global ocean ecosystems. If this discovery proves to be only partially as important as it now seems (and it may not, as is often the case with scientific discoveries and the hypotheses they generate), then our recent large-scale decimation of whales all over the globe may have had a profound, and perhaps irreversible, effect on entire ecosystems that exist thousands of meters below the ocean's surface.

# Chapter 2

# The Gulf of Mexico

## *Physical Description and Human Impacts*

THE GULF OF MEXICO is a semi-enclosed embayment of the Atlantic Ocean bounded by the United States, Mexico, and Cuba (Figure 1). It covers about 1.5 million km$^2$ (0.58 million mi$^2$), opening to the Atlantic Ocean at the narrow Straits of Florida (ca. 860 m or 2820 ft. deep) in the east and to the Caribbean Sea through the deeper (2000 m or 6,560 ft.) Yucatan Channel in the southeast. The Gulf and adjacent Caribbean Sea have been referred to as the American Mediterranean, because both the Gulf of Mexico and the Mediterranean Sea are enclosed almost entirely by land masses and have about the same maximum depth (ca. 5 km or 3.1 mi.). Because the two are so enclosed and adjoin the two most industrially developed continents on earth, North America and Europe, it can be expected that some of the same problems of industrial activity and pollution affect both bodies of water. Some of these issues will be addressed later in the book.

The Gulf of Mexico has a particularly wide continental shelf ringing most of it, with greater than 200-m depths (656 ft.) within 50 km (31 mi.) of shore only off the delta of the Mississippi River (extreme southwest of its extent into the Gulf of Mexico) and at its two mouths (including deep water off the western edge of Cuba). Wide, shallow banks exist along its eastern edge near Florida in the northwest (the Texas-Louisiana shelf) and the south (the Campeche Bank of Mexico). Continental shelf waters less than 200 m (656 ft.) deep comprise about 35 percent of the Gulf surface; the continental slope (200–3000 m, 656–9840 ft.) covers an additional 40 percent; only the remaining 25 percent is composed of truly oceanic depths, mainly in the mid-western part called the Sigsbee Abyssal Plain.

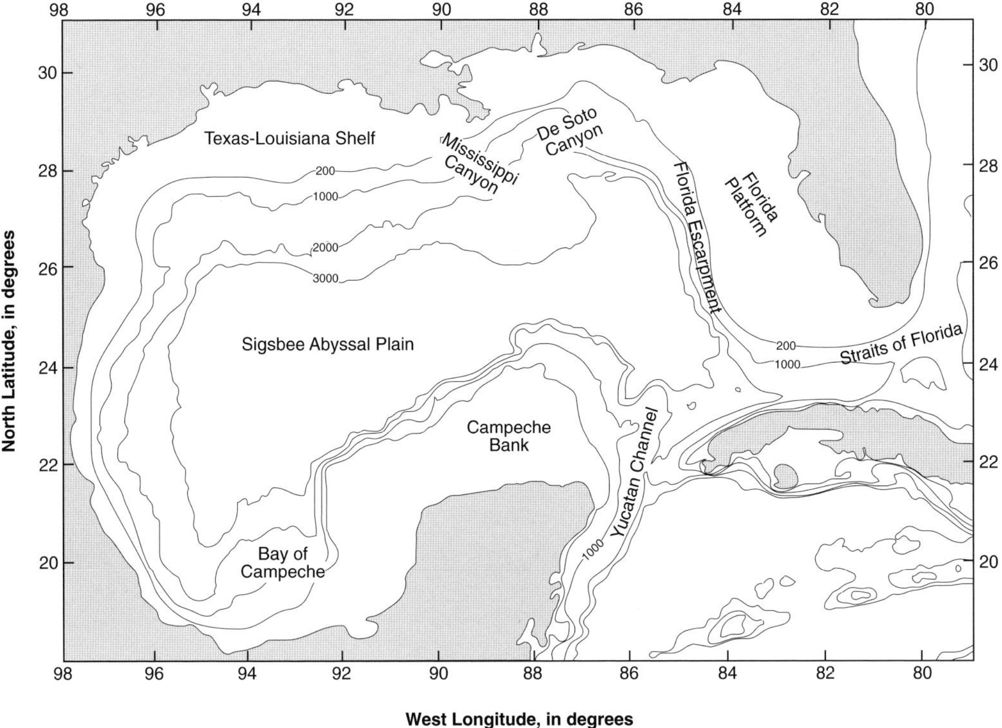


*Figure 1. The Gulf of Mexico, showing major features mentioned in the text. Depths are in meters; multiply these by 3.28 to obtain feet. Two degrees latitude equals 120 nautical miles, or 222 kilometers. Courtesy Cartographic Service Unit, Department of Geography, Texas A&M University*

The wide continental shelf areas of the Gulf of Mexico are smooth and gently sloping. The continental slope, in contrast, is steeply sloping, cut by deep canyons, and dotted with knolls and banks that mark the crests of salt domes, particularly in the western half. At the mouth of the Mississippi River on the continental slope is an underwater extension of the Mississippi River Delta projecting southeastward from the river mouth known as the Mississippi fan. The wide continental shelves of the western Gulf are composed mainly of terriginous sediments—sand, silt, and clay—that originate on land and are transported down rivers. The Florida platform off Florida to the east and the Campeche Bank off the Yucatan peninsula in the south consist of calcareous sediments formed as ancient reefs during the Cretaceous period when the sea level was much lower. The Sigsbee Abyssal Plain is essentially a vast plain of calcareous sediments that originated from shells secreted by marine organisms which drifted down to the sea floor as they died.



The Gulf of Mexico is in a semitropic maritime area. During summer and early fall, the air is moist and warm at an average of about 26°C (77°F) in mid-Gulf. Hot, humid tropical air comes out of the south and southeast in summer. At this time, unstable air systems develop in the tropics and can form into hurricanes in the Gulf, or find their way into the Gulf and wreak disasters when they hit land. In winter, southern winds are weaker, and are often forced back by southward dips of the jet stream—cold fronts to those who live along the shores of the northwest Gulf—and temperatures can drop precipitously from warm-humid to near freezing (and below) in the nearshore areas of the Gulf within one hour or less. The average mid-Gulf temperature in winter is 13°C (55.4°F). The transition periods of April in spring and November in autumn are about 20°C (68°F).

Tides are variable in the Gulf of Mexico due to its enclosed nature and shallow depth. Overall, vertical tide ranges are 0.3–1.2 m (1–4 ft.), depending on location and time of year. Also depending on location, tides may be diurnal (one high and one low per day) or semidiurnal (two highs and two lows per day), or some intermediate stage between the two.

The coastline of the Gulf of Mexico encompasses a wide variety of ecosystems and habitats. In the west are the hypersaline lagoons of northern Mexico and south Texas. Salt marshes line the coast from Texas to Florida in the northern Gulf and occur (less frequently) from the U.S.-Mexico border in the southwest to the Yucatan peninsula in the southeast. Mangrove forests are found in central and southern Florida, the northeastern Gulf of Mexico along the Yucatan, Campeche, Tabasco, and some of the Veracruz coasts of Mexico, and off Cuba. Sea grass beds extend from Florida to the Mississippi River in the east and along the south Texas coastline in the western Gulf; extensive sea grass beds surround the Yucatan peninsula and Campeche coasts to the south. Coral reefs are scattered off western Florida in the east and occur on isolated banks off Texas and Louisiana to the north. Coral reefs are much more common in the south, off the Yucatan peninsula and elsewhere in Mexico. This wide variety of habitats results in enormous productivity in the Gulf of Mexico that provides feeding, spawning, and nursery grounds for nearly 50 species of commercially harvested fish and shellfish.

The major circulation driving the Gulf of Mexico current patterns consists of the warm Yucatan Current coming out of the Caribbean Sea from the south. It loops northward and then eastward, exiting as the Florida Current between Florida and Cuba. In its entirety, this clockwise water-mass movement is termed the Loop Current (Figure 2). It is especially strong in the summer when southeasterly tropical surface waters help drive the Gulf surface waters to the north, resulting in seasonal incursions of the Loop Current onto the West Florida shelf. This wind-driven current "piles up" water at the narrow Yucatan Channel and pushes more water into the Gulf of Mexico than can exit seasonally at the shallower Straits of Florida. As a result, the Gulf maintains a sea level averaging 10 cm (4 in.) higher than the adjacent ocean. The sea level at Cedar Key on the western coast of Florida, for example, is persistently 19 cm (7.5 in.) higher than at St. Augustine, Florida, its approximate latitudinal east coast counterpart on the Atlantic Ocean.



As it meanders northward into the Gulf, extensions of the Loop Current are periodically pinched off and form great (up to 250 km, or 155 miles, in diameter) circular, clockwise-rotating eddies of warm water that drift slowly to the west; there, they dissipate much later off northern Mexico or south Texas after colliding with the continental slope in the so-called "eddy graveyard." In the western Gulf, a clockwise gyre is restricted to the upper 500 m (1640 ft.) of water. There remains some question as to whether it is driven by Loop Current eddies drifting westward or by wind curl stress, or both. Because the giant eddy of warm water adjoining cooler seas moves in a circular flow, water masses can be pulled out of depth, and smaller counterclockwise "cold-core" eddies form at the periphery of the larger warm-core zones. These edges of changing temperatures result in the mixing of local water masses, and can be very biologically productive as nutrient-rich cold waters from below are mixed with waters near the surface in which photosynthesis can take place. These eddies can also pull nutrient-rich waters from the continental shelf to mix with waters of the

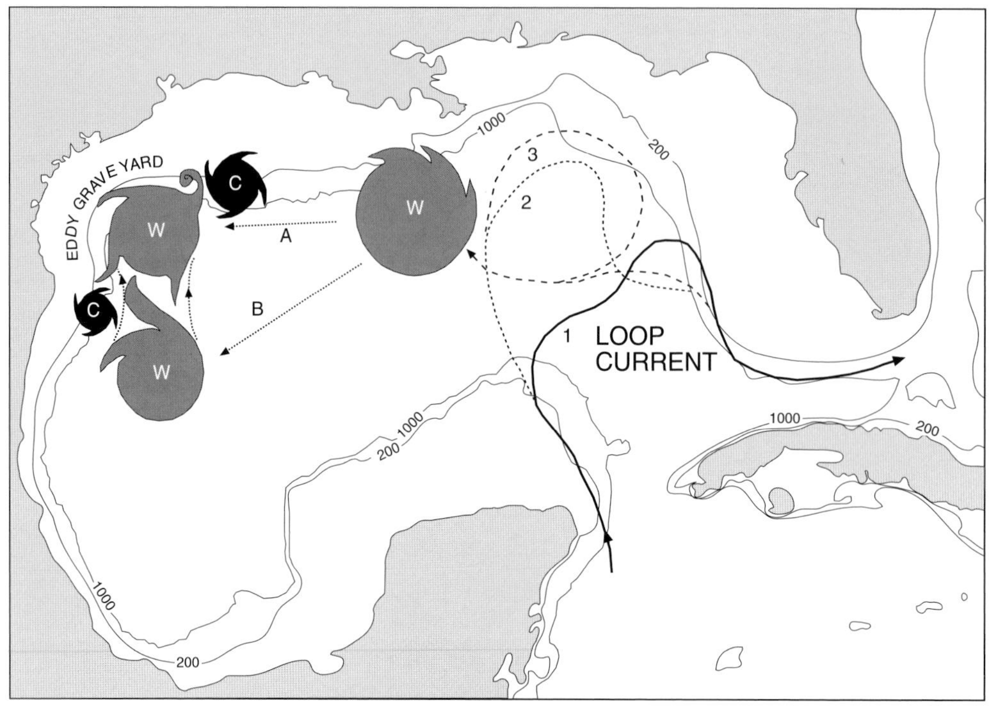


*Figure 2. Schematic of the life cycle of a Loop Current anticyclonic or warm-core eddy (W) as it is pinched off or separated from the Loop Current, and its possible paths through the western Gulf of Mexico. A time series representation of the Loop Current shows it in three possible positions (1–3). The third position represents the most northerly intrusion into the Gulf (this event usually occurs in summer), at which time an anticyclonic eddy may pinch off from the Loop Current. After its formation, the warm-core eddy may follow one of two paths: a westerly (A) or southwesterly (B) path. Cyclonic or cold-core eddies (C) are frequently associated with a warm-core eddy. Regardless of whether an eddy follows path A or B, anticyclonic eddies spin down or fade away in an area of the northwestern Gulf known as the eddy graveyard. This is due to loss of vorticity from colliding with the continental margin. Courtesy Cartographic Service Unit, Department of Geography, Texas A&M University, modified from a figure by F. M. Vukovich and B. W. Crissman*

open Gulf. Such productive areas might be expected to be particularly rich in marine mammals, and—as we shall see in this book—there are some intriguing new data regarding the distribution of certain marine mammals in correlation with warm- or cold-core rings.

While the Loop Current impacts Gulf of Mexico waters from the south, the Mississippi River does so from the north. It is the largest river in North America, and its freshwater effect can reach as far as Corpus Christi in south Texas to the southwest and the Straits of Florida to the east, depending on prevailing winds and nearshore surface currents during and after the major river discharge in spring. About two-thirds of all water from the U.S. mainland drains into the Gulf (an area of nearly 6 million $km^2$, or 2.32 million $mi^2$); most of this comes from the mighty Mississippi and adjacent Atchafalaya River system. This drainage area includes most of the eastern United States as far west as the Rocky Mountains and extends as far north as the southern portions of Canada. Water from one-half of Mexico's land area (ca. 1 million $km^2$, or 0.39 million $mi^2$) drains into the Gulf as well, but from smaller rivers, all the way from the Rio Grande (called the Rio Bravo in Mexico) at the U.S.-Mexico border in the north to the Yucatan peninsula in the southeast. These more diffuse inputs into the Gulf represent runoff mainly from the central Mexican mountain chain (especially between the cities of Tampico to the north and Coatzacoalcos to the south) that extend inland as far as parts of Guatemala. It is the Mississippi River system, however, that contributes about 73 percent of the total freshwater input into the Gulf of Mexico.

Nonetheless, diversion of freshwater inflow into the Gulf of Mexico due to damming and canalization for agricultural and municipal use as well as flood control, has drastically changed sedimentation patterns along adjacent shorelines, and has resulted in accelerated erosion and subsidence. Freshwater inflow into the Gulf varies greatly with season. The highest freshwater inflow is during the spring floods from March to May. By the time it reaches the Gulf, this water has accumulated nearly 6 million $km^2$ (2.32 million $mi^2$) of agricultural, municipal, and industrial wastes. This massive influx of nutrients results in huge blooms of phytoplankton, some of which are toxic to fish, birds, and mammals. The downside of this huge influx of nutrients is termed eutrophication. In this process, the nutrients allow massive blooms of phytoplankton which then die; the bacteria that recycle their remains consume all of the available oxygen in the water column, causing hypoxic ($<2$ mg/L of dissolved oxygen) or even anoxic (no oxygen) conditions. This results in a large "dead zone" to the southwest of the Mississippi River along the coasts of Louisiana and Texas during summer months (June–August).

The Mississippi and other rivers bring high quantities of sediment, nutrient, and pollutant loads from inland. Because nutrients are high in the east, south, and west of the Mississippi Delta, there is concomitant high phytoplankton production and the staggering fact that nearly 40 percent of the entire U.S. commercial fish catch comes from the Texas-Louisiana shelf. Here too, shrimp trawlers ply their trade in very shallow waters just offshore and in bays and inlets, and bottlenose dolphins—those amazingly ubiquitous inhabitants of the nearshore environment—associate with the trawlers and feed on fish and invertebrates stirred up by the giant bottom-dragging nets.



The northern Gulf of Mexico shelves, both east and west of the Mississippi Delta, as well as parts of the Campeche Bank to the south, were once highly productive tropical marshes and forests more than 100 mya. This giant reservoir of biological matter eventually was covered by rising waters, sedimentation, and global tectonic action. These forces, aided by millions of years of intense pressures, have created incredibly large reservoirs of oil and natural gas. Oil production in the northern Gulf has been a major economic force for the United States; although the boom of production is now past, the flickering lights of giant oil rigs are still evident on the continental shelf (in waters generally <200 m or 656 ft. deep) all along the Texas-Louisiana shelf. As many as 15,000 wells have operated in the northern Gulf of Mexico, with a steady decline in the past 20 years. Still, more than 72 percent of the oil and 97 percent of the natural gas produced offshore in the United States come from the Gulf of Mexico. There are more oil reserves out there, however, and talk persists of expanding into oceanic depths off the continental shelf, into waters south of Pensacola and Panama City, Florida. The possible impact of this development on some marine mammals in the northern Gulf of Mexico will be discussed in chapter 5.

Human impacts such as those resulting from offshore drilling and shipping are amplified by the enclosed nature of the Gulf of Mexico. More than seventy countries traverse the Gulf or exploit its resources in some way. Four of the United States' ten busiest ports are located in the Gulf of Mexico, handling almost 45 percent of U.S. waterborne tonnage. In addition, there are eight Navy home ports in the Gulf of Mexico. Wastes from these intensive shipping activities include bilge washings, sewage, galley wastes, marine debris, and accidental discharges of oil and other hazardous materials.



Do not, however, assume by the aforementioned general description that oceanic circulations, atmospheric movements, productivity, and human effects in the Gulf are well understood or predictable. We know that in summer we can have large, warm-core rings and hurricanes. We know that in winter these tendencies are lessened, and in spring the flow of the Mississippi becomes especially important. We, however, cannot predict particular movements of air and water masses, or specific areas of primary or secondary productivities. We therefore yet cannot predict the movements of marine mammals in this dynamic, continually changing environment. We know that sperm whales, beaked whales, and several other toothed whales stay in deeper waters; thus, deepwater marine mammals such as sperm whales would be expected close to shore in U.S. waters only off the Mississippi Delta and, indeed, research has shown this to be the case. We believe that bottlenose dolphins are creatures of both inland waters and, along with Atlantic spotted dolphins, the continental shelf (as separate populations). We know that warm- and cold-core rings and especially their interfaces (or confluences) may be important for some species of toothed whales in the Gulf of Mexico. More specifically, we have recently found that in the north-central Gulf of Mexico, many cetaceans preferentially use the areas where cold-core rings or eddies occur. These circular patterns are associated with the generally larger, less productive warm-core rings that have budded off the relatively "hot" Loop Current emanating from the Yucatan Channel. Beyond that we know little, and it remains for further oceanography–marine mammal correlations to help us predict where and when a certain species might appear in this marvelously dynamic, constantly changing world.

The Gulf of Mexico has been the site of extensive human activities in the past few decades. In particular, offshore oil and gas exploration and drilling have been major activities over the continental shelf (water depths: <200 m or 656 ft.), largely off Mississippi, Louisiana, and Texas. In 1983, 311 million barrels of oil were obtained from U.S. reserves in the Gulf, about 11 percent of the total U.S. production. An extensive industry surrounds the extraction of oil offshore, characterized by production and pumping platforms, tanker traffic involved in moving the oil, seismic surveys used in finding oil reserves, explosive removal of platforms from expired lease areas, and associated support services (ships and aircraft used to transport personnel and supplies). Recently, technology has been developed to allow exploitation of oil and gas deposits in deeper (>200 m or 656 ft.) waters of the Gulf's continental slope, necessitating information on more oceanic species.

Thus far, the effects of most of these activities on marine mammals of the Gulf of Mexico have not been examined in detail. In particular, it is essential to have some idea of the species composition, distribution, abundance, and especially behavior and ecology of different cetaceans in areas in which oil and gas activities occur, so that any deleterious effects on the cetacean community can be assessed and minimized. While some recent studies have revolutionized our understanding, many more are needed; we also must be particularly inventive in using techniques and resources to learn how marine mammals live, and how they are affected by human activity.

*1. Northern right whale and newborn,* Eubalaena glacialis. *Photograph by Bernd Würsig*

*2. (below) Blue whale, head.* Balaenoptera musculus. *Photograph by Michael Williamson*

*3. Fin whale, head,* Balaenoptera physalus. *Photograph by Kenneth C. Balcomb*

*4. Sei whale, whole body, underwater,* Balaenoptera borealis. *Photograph by James D. Watt*

*5. Bryde's whale, head,* Balaenoptera edeni. *Photograph by Robert Pitman*

*6. Minke whale, head,* Balaenoptera acutorostrata. *Photograph Tom Jefferson*

*7. Humpback whale, body and tail,* Megaptera novaeangliae. *Photograph by Tom Jefferson*

*8. Sperm whale, back and fin, blowing,* Physeter macrocephalus. *Photograph by Bernd Würsig*

*9. Pygmy sperm whale, body and fin,* Kogia breviceps. *Photograph by Robert Pitman*

*10. Dwarf sperm whale, leaping,* Kogia simus. *Photograph by Robert Pitman*

*11. Cuvier's beaked whale, body and head, distant,* Ziphius cavirostris. *Photograph by Robert Pitman*

*12. Blainville's beaked whale, head,* Mesoplodon densirostris. *Photograph by S. Jonathan Stern*

*13. Gervais' beaked whale, full head,* Mesoplodon europaeus. *Photograph by David Weller*

*14. Killer whale, head and body, fin,* Orcinus orca. *Photograph by Robert Pitman*

*15. Short-finned pilot whales, body and fin,* Globicephala macrorhynchus. *Photograph by Bernd Würsig*

*16. Long-finned pilot whale pod,* Globicephala melas. *Photograph by Robert Pitman*

*17. False killer whale, body and fin, above,* Pseudorca crassidens. *Photograph by Robert Pitman*

*18. Pygmy killer whale, most of head and body, underwater,* Feresa attenuata. *Photograph by Robert Pitman*

*19. Melon-headed whale, most of body, underwater,* Peponocephala electra. *Photograph by Keith Mullin*

*20. Rough-toothed dolphin, body, underwater,* Steno bredanensis. *Photograph by Bernd Würsig*

21. *Risso's dolphin, body, fin,* Grampus griseus. *Photograph by Tom Jefferson*

22. *Bottlenose dolphin, leaping,* Tursiops truncatus. *Photograph by Dagmar C. Fertl*

*23. Bottlenose, full, underwater,* Tursiops truncatus. *Photograph by Bernd Würsig*

*24. Pantropical spotted dolphin, single, leaping,* Stenella attenuata. *Photograph by Carol L. Roden*

25. *Atlantic spotted dolphin, head and body, underwater,* Stenella frontalis. *Photograph by Bernd Würsig*

26. *Spinner dolphin, full leap,* Stenella longirostris. *Photograph by Carol L. Roden*

*27. Clymene dolphin, full leap,* Stenella clymene. *Photograph by Carol L. Roden*

*28. Striped dolphin, full, leaping,* Stenella coeruleoalba. *Photograph by Carol L. Roden*

*29. Short-beaked common dolphin, leaping,* Delphinus delphis. *Photograph by Robert Pitman*

*30. Long-beaked common dolphin, leaping,* Delphinus capensis. *Photograph by Robert Pitman*

*31. Fraser's dolphin, body and fin, leaping,* Lagenodelphis hosei. *Photograph by Scott R. Benson*

*32. West Indian manatee,* Trichechus manatus. *Photograph by Tamara Miculka*

# Chapter 3

# Synopsis and Keys to the Marine Mammals of the Gulf of Mexico

THIS CHAPTER INCLUDES a synopsis of the overall marine mammal fauna in the Gulf of Mexico including a taxonomic checklist and brief characterization of status, two species identification keys for cetaceans (one based on external appearance and the other on features of the teeth, skull, and skeleton), and a brief history of marine mammal research in the region. A checklist with scientific and common names for marine mammals in the entire world is presented in the Appendix. Finally, illustrations of the external appearance of each species and the skull of cetacean species have been included to assist the interested reader and observer with species identifications.

Thirty-one species (28 cetaceans, 2 pinnipeds, and 1 manatee) of marine mammals are known from the Gulf of Mexico. At least three additional species (the long-finned pilot whale and the short-beaked common and the long-beaked common dolphins) have been recorded so close to the boundaries of the region that they may be found eventually in the area. Of the two pinnipeds, one (the Caribbean monk seal) is now extinct, and the other (the California sea lion) was introduced and occurred only in the feral condition during the past. The West Indian manatee is the only sirenian in the region and is listed as an endangered species in threat of extinction.

The cetacean fauna includes nine cosmopolitan species that occur in most major oceans, and for the most part are eurythermic with a broad range of temperature tolerances. These are the minke, sperm, pygmy sperm, dwarf sperm, Cuvier's beaked, Blainville's beaked, and killer whales, and Risso's and bottlenose dolphins.

Four species (Sowerby's beaked and Gervais's beaked whales, and the Atlantic spotted and Clymene dolphins) have a distribution confined to the Atlantic Ocean. Of these, one species (Sowerby's beaked whale) is of extralimital occurrence in the Gulf.

Ten cetaceans have distributions peculiar to tropical, warm temperature waters of both hemispheres and may be considered warm-stenothermal forms. These include the Bryde's, short-finned pilot, false killer, pygmy killer, and melon-headed whales, and the rough-toothed, pantropical spotted, spinner, striped, and Fraser's dolphins.

Five species (the right, blue, fin, sei, and humpback whales) have disjunct bipolar (anti-tropical) distributions and are regarded as cold-stenothermal forms based on where they feed. The humpback whale, however, enters tropical waters to breed.

Several migratory species occur in the Gulf of Mexico, including several species of rorquals (the blue, fin, sei, and minke whales), the humpback, and the northern right whales. These species generally travel between a breeding zone in which they do not eat and a feeding zone of high productivity in cooler waters. The sperm whale is known to be migratory in other parts of the world, but female groups appear to be year-round residents in the Gulf of Mexico.

Some of the larger whales that occur in the Gulf have been placed on the Endangered Species List of the U.S. Fish and Wildlife Service Register. These include the sei, fin, blue, and right whales. Only two, the blue and right whales, are considered endangered on a worldwide basis. The sperm whale also is listed as endangered, but recent information suggests that it is the most common large whale in the Gulf of Mexico.

No dolphins or other odontocetes in the Gulf are considered endangered at the species level, although the spinner dolphin has received considerable attention as a locally threatened species in the eastern Pacific where large numbers have been killed incidentally in the tuna purse seine fisheries.

No mention of marine mammals of the Gulf would be complete without a description of legal protective measures. Fortunately, these mammals are protected by the Marine Mammal Protection Act (MMPA) of 1972 which states that it is illegal to hunt or otherwise do damage to the animals. The actual enforcement is

*1. Texas A&M University researcher Kathleen Dudzinski free diving with Atlantic spotted dolphins to photographically identify individuals, ascertain relative age by spot patterns, and determine gender from photos of the genital area. Courtesy Bernd Würsig*

under the jurisdiction of the U.S. National Marine Fisheries Service for whales, dolphins, porpoises, and pinnipeds, and the U.S. Fish and Wildlife Service is responsible for walrus, polar bears, sea otters, and manatees. In short, it is illegal to harm marine mammals, even to cause behavioral changes. U.S. regulation forbids harassing marine mammals by approaching them too closely, and prohibits feeding them. Some limited accidental fisheries bycatch is allowed for non-endangered species. Species considered endangered, such as the sperm whale and manatee in the Gulf waters, are further protected by the U.S. Endangered Species Act (ESA). In Mexico, capture, exploitation, and trade of marine mammals or their body parts are forbidden without special permits. Marine mammal protection is coordinated by the Secretaría del Medio Ambiente, Recursos Naturales y Pesca, but there is no national law equivalent to the MMPA. We do not know the status and legal protective measures of marine mammals in Cuban waters.

## Checklist

The marine mammals of the Gulf of Mexico are listed below, including the scientific name, common name, and status of each species. The checklist includes those species recorded from the

Gulf as well as three species known from nearby waters that may be documented in the region following additional work and investigation.

Population status in the Gulf is summarized according to the following categories:



COMMON: A common species is one that is abundant wherever it occurs in the region. Most common species are widely distributed over the area.

UNCOMMON: An uncommon species may or may not be widely distributed but does not occur in large numbers. Uncommon species are not necessarily rare or endangered.

RARE: A rare species is one that is present in such small numbers throughout the region that it is seldom seen. Although not threatened with extinction, a rare species may become endangered if conditions in its environment change.

POSSIBLE: A possible species is one that has not been recorded from the Gulf, but is known from adjacent areas and eventually could be documented from the region.

EXTINCT: An extinct species is one that no longer exists.

EXTRALIMITAL: An extralimital species is known on the basis of only a few records that probably resulted from unusual wanderings of animals into the region.

INTRODUCED: An introduced species is one that does not occur naturally in the region, but was introduced and may occur in the feral condition.

The distributional status of each species in the checklist has been summarized according to whether the species is cosmopolitan or worldwide in distribution (cosmopolitan), endemic and known only from the Atlantic Ocean (endemic to Atlantic), a warm-stenothermal species with a tropical distribution (tropical), or a bipolar species with an anti-tropical distribution (anti-tropical).

Finally, the basis for documenting the species in the Gulf is provided according to whether the record is based on the verified documentation of a stranded animal, an observation made from a ship or plane by a scientific expert or reputable observer of whales and dolphins, or both types of documentation.

Order Cetacea

Suborder Mysticeti, baleen whales

Family Balaenidae, right whales

*Eubalaena glacialis,* northern right whale: extralimital, cosmopolitan, anti-tropical, stranding and observation.

Family Balaenopteridae, rorquals

*Balaenoptera musculus,* blue whale: extralimital, cosmopolitan, stranding.

*Balaenoptera physalus,* fin whale: rare, cosmopolitan, stranding and observation.

*Balaenoptera borealis,* sei whale: rare, anti-tropical, stranding.

*Balaenoptera edeni,* Bryde's whale: uncommon, tropical, stranding and observation.

*Balaenoptera acutorostrata,* minke whale: rare, cosmopolitan, stranding and observation.

*Megaptera novaeangliae,* humpback whale: rare, cosmopolitan, stranding and observation.

Suborder Odontoceti, toothed whales and dolphins

Family Physeteridae, sperm whale

*Physeter macrocephalus,* sperm whale: common, cosmopolitan, stranding and observation.

Family Kogiidae, pygmy and dwarf sperm whales

*Kogia breviceps,* pygmy sperm whale: uncommon, cosmopolitan, stranding and observation.

*Kogia simus,* dwarf sperm whale: uncommon, cosmopolitan, stranding and observation.

Family Ziphiidae, beaked whales

*Ziphius cavirostris,* Cuvier's beaked whale: rare, cosmopolitan, stranding and observation.

*Mesoplodon densirostris,* Blainville's beaked whale: rare, cosmopolitan, stranding.

*Mesoplodon bidens,* Sowerby's beaked whale: extralimital, endemic to Atlantic, stranding.

*Mesoplodon europaeus,* Gervais' beaked whale: uncommon, endemic to Atlantic, stranding.

Family Delphinidae, ocean dolphins

*Orcinus orca,* killer whale: uncommon,

cosmopolitan, stranding and observation.

*Globicephala macrorhynchus,* short-finned pilot whale: common, tropical, stranding and observation.

*Globicephala melas,* long-finned pilot whale: possible, anti-tropical, no documented records from Gulf.

*Pseudorca crassidens,* false killer whale: uncommon, tropical, stranding and observation.

*Feresa attenuata,* pygmy killer whale: uncommon, tropical, stranding and observation.

*Peponocephala electra,* melon-headed whale: common, tropical, stranding and observation.

*Steno bredanensis,* rough-toothed dolphin: common, tropical, stranding and observation.

*Grampus griseus,* Risso's dolphin: common, cosmopolitan, stranding and observation.

*Tursiops truncatus,* bottlenose dolphin: common, cosmopolitan, stranding and observation.

*Stenella attenuata,* pantropical spotted dolphin: common, tropical, stranding and observation.

*Stenella frontalis,* Atlantic spotted dolphin: common, endemic to Atlantic, stranding and observation.

*Stenella longirostris,* spinner dolphin: common, tropical, stranding and observation.

*Stenella clymene,* Clymene dolphin: common, endemic to Atlantic, stranding and observation.

*Stenella coeruleoalba,* striped dolphin: common, tropical, stranding and observation.

*Delphinus delphis,* short-beaked common dolphin: possible, cosmopolitan, no documented records from Gulf.

*Delphinus capensis,* long-beaked common dolphin: possible, tropical, no documented records from Gulf.

*Lagenodelphis hosei,* Fraser's dolphin: common, tropical, stranding and observation.

Order Sirenia, the dugong and manatees

Family Trichechidae, manatees

*Trichechus manatus,* West Indian manatee: common, endemic to Atlantic, stranding and observation.

Order Carnivora
Suborder Pinnipedia, seals and sea lions
Family Phocidae, true seals
*Monachus tropicalis,* Caribbean monk seal: extinct.
Family Otariidae, sea lions
*Zalophus californianus,* California sea lion: introduced.



## Identification Keys for Cetaceans

Two keys have been prepared for the 31 species of cetaceans that are known to occur (28) or are expected to occur (3) in the Gulf of Mexico. These keys are presented for those not thoroughly familiar with the diagnostic characters of Gulf cetaceans. It must be cautioned, however, that identification should not be based on a single character, but on a suite of features. For that reason, when the key resolves to a species, a list of all the appropriate diagnostic characters is presented rather than only those needed to separate it from the other member of the dichotomy. In this way, an observer can verify that the identification is correct, thereby avoiding errors based on specimens not fitting the typical pattern for one character.

### KEY TO IDENTIFICATION BASED ON EXTERNAL APPEARANCE

The first key is based on external characters in which measurements and tooth–baleen plate counts can be taken, and detailed observations can be made of color patterns and body features that normally are not visible on animals observed at sea. (This key will be of limited use for sightings at sea.) This key primarily is intended to reflect diagnostic features documented in adult specimens, and in some cases may not allow identification of subadults. It also must be cautioned that some groups, such as the beaked whales of the genus *Mesoplodon,* are poorly known, and this key probably will not be adequate to accurately identify these species. Finally, a caveat regarding the use of color patterns for identification: the darkening of the skin that occurs postmortem in cetaceans can result in the disappearance of some of the more subtle color patterns within hours of death, even when the specimen is kept cool. This should be kept in mind when using subtle features of color pattern for identification. Some of the features described in this key are depicted in the line drawings of the various species that follow this section.



1a. Double blowhole; teeth absent; baleen plates suspended from upper jaw. Go to 2.

b. Single blowhole; teeth present (although sometimes not protruding from gums); no baleen plates. Go to 8.

2a. No creases or pleats on chin or throat; no dorsal fin or hump; upper jaw and mouth line strongly arched when viewed from side and very narrow from top; callosities (roughened areas of skin to which whale lice attach) on head; 200–270 long, narrow (<2.8 m, 9.2 ft.) baleen plates/side, black with fine black fringes; body black, often with white ventral blotches; maximum body length 17 m, 56 ft.

NORTHERN RIGHT WHALE, *Eubalaena glacialis*

b. Long ventral pleats; dorsal fin present; upper jaw relatively flat when viewed from side and broad from top. Go to 3.

3a. Flippers ¼–⅓ of body length, with knobs on leading edge; flukes with irregular trailing edges; <35 broad, conspicuous ventral pleats, longest extending at least to navel; top of head covered with knobs, 1 prominent cluster of knobs at tip of lower jaw; 270–400 black to olive brown baleen plates with gray bristles/side (<80 cm, 31.5 in. long); dorsal fin usually atop a hump; maximum body length 16 m, 52.5 ft.

HUMPBACK WHALE, *Megaptera novaeangliae*

b. Flippers <⅕ of body length, lacking knobs; 32–100 fine ventral pleats; head lacking knobs; flukes more or less smooth on trailing edges. Go to 4.

4a. Ventral pleats end before navel. Go to 5.

b. Ventral pleats extend to or beyond navel. Go to 6.

5a. Ventral pleats 50–70, longest ending before navel (often ending between flippers); 231–285 baleen plates with coarse bristles/side (<21 cm, 8.3 in. long), mostly white or yellowish white (sometimes with dark margin along outer edge); often conspicuous white bands on upper surface of flippers;

from above, head sharply pointed; maximum body length 10 m.

MINKE WHALE, *Balaenoptera acutorostrata*

b. Ventral pleats 32–60, longest ending past flippers, but well short of navel; 219–402 pairs of black baleen plates with many fine whitish bristles (<80 cm, 31.5 in. long); flippers all dark; snout turned slightly downward at tip when viewed from side; maximum body length 16 m, 52.5 ft.

SEI WHALE, *Balaenoptera borealis*

6a. Three conspicuous ridges on rostrum; 40–70 ventral pleats extending to umbilicus; prominent dorsal fin rising at steep angle from back; 250–370 slate gray baleen plates/side with white to light gray fringes; head coloration symmetrical; maximum body length 16 m, 52.5 ft.

BRYDE'S WHALE, *Balaenoptera edeni*

b. Only 1 prominent ridge on snout (although slight accessory ridges may be present); 55–100 ventral pleats. Go to 7.

7a. Head broad and almost U-shaped from above; dorsal fin very small (ca. 1 percent of body length) and set far back on body; 270–395 black baleen plates with black bristles/side (all 3 sides of each plate roughly equal in length); head coloration symmetrical; body mottled gray, with white lower surface of flippers; maximum body length 30 m, 98.4 ft.

BLUE WHALE, *Balaenoptera musculus*

b. From above, head more V-shaped and pointed at tip; dorsal fin about 2.5 percent of body length, rising at shallow angle from back; 260–480 gray baleen plates with white streaks/side (front ⅓ of baleen on right side all white); head coloration asymmetrical (left side gray, much of right side white); back dark with light streaks and chevron; belly white; maximum body length 24 m, 78.7 ft.

FIN WHALE, *Balaenoptera physalus*

8a. Upper jaw extending well past lower jaw; lower jaw very narrow. Go to 9.

b. Upper jaw not extending much or at all past lower jaw; lower and upper jaws about equal in width (beaked whale or delphinid). Go to 11.

9a. Body black to charcoal gray with white lips and inside of mouth; head squarish and large (20–30 percent of body length); short creases on throat; S-shaped blowhole on left side at front of head; low, rounded dorsal "hump" followed by series of bumps along midline; 18–25 heavy, peglike teeth/side of lower jaw that fit into sockets in upper jaw; body length 4–18 m.

SPERM WHALE, *Physeter macrocephalus*

b. Body <4 m, 13.1 ft.; head <15 percent of body length; blowhole set back from front of head; prominent dorsal fin; 8–16 long, thin, sharply pointed teeth in each side of lower jaw that fit into upper jaw sockets. (Note: Both species of *Kogia* generally are difficult for nonexperts to distinguish.) Go to 10.

10a. Throat creases generally absent; short, rounded dorsal fin (<5 percent of body length) located in last ⅓ of back; distance from tip of rostrum to blowhole >10.3 percent of total length; 12–16 (rarely 10–11) sharp teeth in each half of lower jaw; maximum body length 3.4 m, 11.2 ft.

PYGMY SPERM WHALE, *Kogia breviceps*

b. Inconspicuous throat creases; dorsal fin relatively tall (>5 percent of body length) and "dolphin-like," located near middle of back; distance from tip of rostrum to blowhole <10.2 percent of total length; 8–11 (rarely 12–13) long, fanglike teeth/side of lower jaw (sometimes 1–3 in each half of upper jaw); maximum body length 2.7 m, 8.9 ft.

DWARF SPERM WHALE, *Kogia simus*

11a. Two conspicuous creases on throat forming forward-pointing V; notch between flukes usually absent or indistinct; dorsal fin relatively short and set well behind mid-body. Go to 12.

b. No conspicuous creases on throat; prominent median notch in flukes; dorsal fin near middle of back or in forward ⅓. Go to 15.

12a. Short beak indistinct; head small relative to body size; forehead slightly concave in front of blowhole; 1 pair of conical teeth directed forward and upward at tip of lower jaw (exposed only in adult males); mouth line upturned at gape; head often light colored; maximum body length 7.5 m, 24.6 ft.

CUVIER'S BEAKED WHALE, *Ziphius cavirostris*

b. Usually 1 pair of flattened teeth well behind tip of lower jaw (erupted only in adult males); head small; prominent beak with forehead rising at shallow angle; sometimes flippers fit into depressions on the body; scratches and scars common on some animals; maximum body length 5.5 m, 19.7 ft.

*Mesoplodon* sp. (Note: All except adult males will require museum preparation for identification, and species other than those listed below may be present in the Gulf.) Go to 13.



13a. Lower jaw usually light in color; tusks of males very large, located on bony prominences near corners of mouth (extend above level of upper jaw), and oriented slightly forward; lower jaw massive (particularly in adult males) with high arching contour; forehead has concavity in front of blowhole; maximum length 5 m, 16.4 ft. (Note: Females and subadults require museum preparation for identification.)
BLAINVILLE'S BEAKED WHALE, *Mesoplodon densirostris*

b. Mouth line relatively straight or only slightly arched; tusks of adult males not on arches; maximum body length 5.5 m, 18.0 ft. Go to 14.

14a. Two small flattened teeth near front of lower jaw of males; beak short to moderate in length. (Note: Females and subadults require museum preparation for identification.)
GERVAIS' BEAKED WHALE, *Mesoplodon europaeus*

b. Teeth of adult males protrude outside mouth in middle of lower jaw; vestigial teeth sometimes present in both jaws; beak long and slender. (Note: Females and subadults require museum preparation for identification.)
SOWERBY'S BEAKED WHALE, *Mesoplodon bidens*

15a. Head blunt with no prominent beak. Go to 16.

b. Head with prominent beak. Go to 21.

16a. Teeth (2–7 pairs) at front of lower jaw only (rarely 1–2 pairs in upper jaw), but teeth may be absent or extensively worn; forehead blunt with shallow, vertical crease; dorsal fin tall and dark; body gray to white, generally covered with scratches and splotches in adults; flippers long and sickle-shaped; maximum body length 4 m, 13.1 ft.
RISSO'S DOLPHIN, *Grampus griseus*

b. Teeth (>7 pairs) in both upper and lower jaws; forehead without vertical median crease; predominant color black or dark gray. Go to 17.

17a. Flippers large and paddle-shaped; dorsal fin tall and erect (<0.9 m in females and 1.8 m in males); striking black-and-white coloration with white postocular patches, white lower jaw, white ventrolateral field, and light gray saddle patch behind dorsal fin; 10–12 large (<2.5 cm or 1 in. in diameter) oval teeth/row; maximum body length 10 m, 32.8 ft.

KILLER WHALE, *Orcinus orca*

b. Flippers long and slender with pointed or blunt tips. Go to 18.

18a. Dorsal fin low and broad based, located on forward ⅓ of back; head bulbous; body black to dark gray with light anchor-shaped patch on belly and often light gray saddle behind dorsal fin; often light streak above and behind each eye; deepened tailstock; long, sickle-shaped flippers; 7–13 pairs of teeth in front half only of each jaw. Go to 19.

b. Dorsal fin near middle of back. Go to 20.

19a. Flipper length 18–27 percent of body length, with prominent "elbow"; 8–13 teeth/row; maximum body length 6.3 m, 20.7 ft. (Note: Distribution is limited to cold temperate regions of the North Atlantic and southern hemisphere.)

LONG-FINNED PILOT WHALE, *Globicephala melas*

b. Flipper length 16–22 percent of body length; 7–9 pairs of teeth/row; maximum body length 6.1 m, 20.0 ft. (Note: Distribution is limited to tropical and warm temperate waters.)

SHORT-FINNED PILOT WHALE, *Globicephala macrorhynchus*

20a. Flippers with distinct humps on leading edge; body predominantly black; no beak; 7–12 large, stout teeth in each half of both jaws, circular in cross section; maximum body length 6 m, 19.7 ft.

FALSE KILLER WHALE, *Pseudorca crassidens*

b. Body black or dark gray with white lips; white to light gray patch on belly; flippers lack humps on leading edges; 8–25 teeth/row. Go to 21.

21a. Less than 15 stout teeth in each half of both jaws; flippers slightly rounded at tip; distinct dorsal cape; head rounded from above and side; maximum body length 2.6 m, 8.5 ft.

PYGMY KILLER WHALE, *Feresa attenuata*

b. More than 15 slender teeth/side in each jaw; flippers sharply pointed at tip; face with dark "mask" (often not visible if specimen is not extremely fresh); light stripe on melon that widens from blowhole to rostrum tip; faint cape that dips low below dorsal fin; head blunt and triangular from above (extremely short, indistinct beak may be present in younger animals); maximum body length 2.75 m, 9.0 ft.

MELON-HEADED WHALE, *Peponocephala electra*



22a. Head long and conical; beak runs smoothly into forehead with no crease; body dark gray to black above and white below with many scratches and splotches; narrow dorsal cape; flippers very large; 19–28 teeth in each half of both jaws, each with series of shallow, vertical wrinkles; maximum body length 2.8 m, 9.2 ft.

ROUGH-TOOTHED DOLPHIN, *Steno bredanensis*

b. Beak distinct from forehead (set off by a crease). Go to 23.

23a. Flippers, flukes, and dorsal fin relatively small; broad dark stripe from eye to anus area on some animals; dorsal fin only slightly recurved; body stocky; extremely short, but well-defined beak (<2.5 percent of body length); deep grooves on palate; 38–44 teeth/row; maximum body length >2.7 m, 8.9 ft.

FRASER'S DOLPHIN, *Lagenodelphis hosei*

b. Beak moderate to long (>3 percent of body length); appendages of normal dolphin proportions. Go to 24.

24a. Body moderately robust; 20–26 teeth in each half of upper jaw, 18–24 in lower jaw (teeth may be extensively worn or missing); moderately long, robust snout set off by distinct crease; color dark to light gray dorsally, fading to white or even pink on belly; dark dorsal cape and light spinal blaze sometimes visible; maximum body length 3.8 m, 12.5 ft.

BOTTLENOSE DOLPHIN, *Tursiops truncatus*

b. Teeth/row >30. Go to 25.

25a. Erect to slightly falcate dorsal fin; dark back and white belly; tan to buff thoracic patch and light gray streaked tail stock form hourglass pattern crossing below dorsal fin; chin to flipper stripe; 41–60 teeth/row; palate with 2 deep longitudinal grooves; maximum body length 2.6 m, 8.5 ft. Go to 26.

b. No hourglass pattern on side; flipper stripe (if present) runs from eye or gape, not chin; palatal grooves shallow (if present). Go to 27.

26a. Body relatively stocky; beak shorter; melon rounded; thoracic patch relatively light; flipper stripe not approaching gape, narrowing ahead of eye; eye patch dark and distinct; light patches often on flippers and dorsal fin; 41–54 teeth/row; maximum body length 2.3 m, 7.5 ft.

SHORT-BEAKED COMMON DOLPHIN, *Delphinus delphis*



b. Body relatively slender; beak longer; melon flatter; thoracic patch not contrasting as strongly with cape; flipper stripe meets lip patch near or just ahead of gape, remains wide ahead of eye; eye patch not as strongly contrasting; light patches on extremities faint if present; 47–60 teeth/row; maximum body length 2.6 m, 8.5 ft.

LONG-BEAKED COMMON DOLPHIN, *Delphinus capensis*

27a. Color pattern black to dark gray on back and white on belly, with prominent black stripes from eye to anus and eye to flipper; light gray spinal blaze extending to below dorsal fin (not always visible); 39–55 teeth/row; maximum body length 2.6 m, 8.5 ft.

STRIPED DOLPHIN, *Stenella coeruleoalba*

b. No continuous stripe from eye to anus. (Note: If a stripe from the anus is present, it does not reach the eye.) Go to 28.

28a. Generally, color pattern two-part (dark cape with lighter sides and belly); beak tip light; slight to heavy spotting present on dorsum of adults (on some individuals, spots may be absent); flipper stripe (if present) runs from gape, not eye; no palatal grooves. Go to 29.

b. Color pattern three-part (white belly, light gray sides, dark gray cape); beak tip dark; usually dark line on top of snout; no spotting on dorsum of adults; cape dips only slightly, to lowest point at level of dorsal fin; eye-to-flipper stripe present; shallow palatal grooves sometimes present. Go to 30.

29a. Body moderately robust, dark gray above with white belly; light spinal blaze (sometimes obscured by spots); slight to heavy spotting on adults (spotting may be nearly absent on subadults); 30–42 teeth/row; maximum body length 2.3 m, 7.5 ft.

ATLANTIC SPOTTED DOLPHIN, *Stenella frontalis*

b. Dorsal fin narrow and strongly falcate; dark cape sweeps to lowest point on side in front of dorsal fin; dark gape to flipper stripe; beak tip and lips white; adults generally with light dorsal spotting and gray bellies (spotting sometimes absent);

34–48 teeth in each half of each jaw; maximum body length 2.6 m, 8.5 ft.

PANTROPICAL SPOTTED DOLPHIN, *Stenella attenuata*

30a. Cape dips prominently above eye and below dorsal fin; snout light gray with dark tip, dark lips, and black line from tip to apex of melon; dark "mustache" present on top of beak; beak generally <7 percent of total length; 38–52 teeth/row; maximum body length 2 m, 6.6 ft.

CLYMENE DOLPHIN, *Stenella clymene*

b. Margin of cape relatively straight, only dipping slightly below dorsal fin; "moustache" not present on beak; dorsal fin slightly falcate to erect; beak exceedingly long and slender (generally >7 percent of total length); 45–64 very fine, sharply pointed teeth/row; maximum body length 2.4 m, 7.9 ft.

SPINNER DOLPHIN, *Stenella longirostris*



## KEY TO SKULLS

This second key, based on the cranial and dental features of adults, includes the same species that appear in the preceding key. For purposes of this key, the user will need a cleaned skull and a metric measuring tape or dial calipers capable of measuring the length of bones or teeth. This key is founded mostly on information from other parts of the world, and only a small amount of data are from the Gulf of Mexico. Geographic variation may make it unreliable for identifying skulls from the Gulf in some cases. Many of the features described in this key are depicted in the skull drawings of the various species that are imbedded in this section.

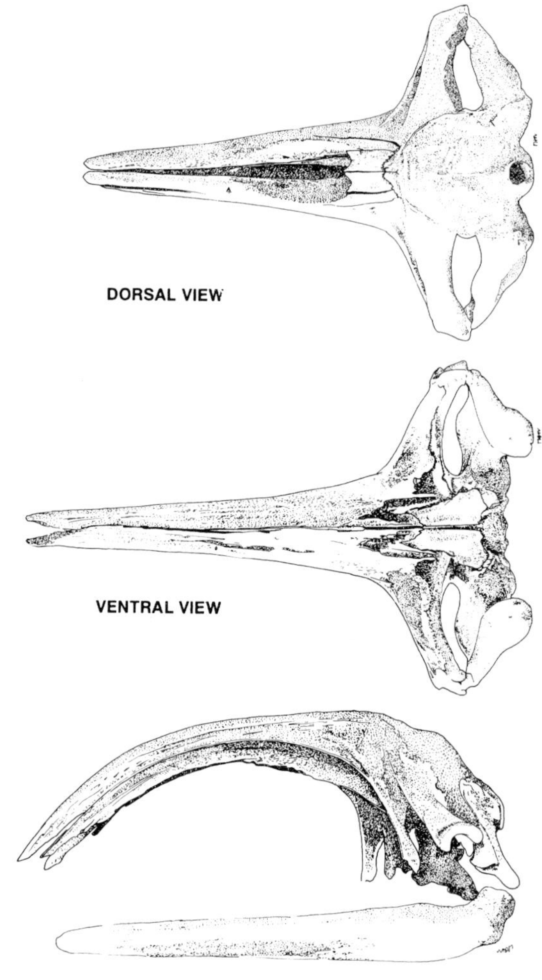


*2. Northern right whale,* Eubalaena glacialis. *Drawn by M. D'Antoni,* FAO Species Identification Guide, Marine Mammals of the World, *Food and Agriculture Organization of the United Nations/United Nations Environment Programme, 1993*

1a. Teeth absent; skull bilaterally symmetrical; lower jaw lacking bony symphysis; size always large (adults >1 m). Go to 2.

b. Teeth present (although they may not emerge from the jawbones in some beaked whales); skull generally asymmetrical; lower jaw possessing bony symphysis; skull generally small (<1.5 m, except in *Physeter*). Go to 6.

2a. Rostrum strongly arched from lateral view; base of rostrum <⅓ cranial width.

NORTHERN RIGHT WHALE, *Eubalaena glacialis*

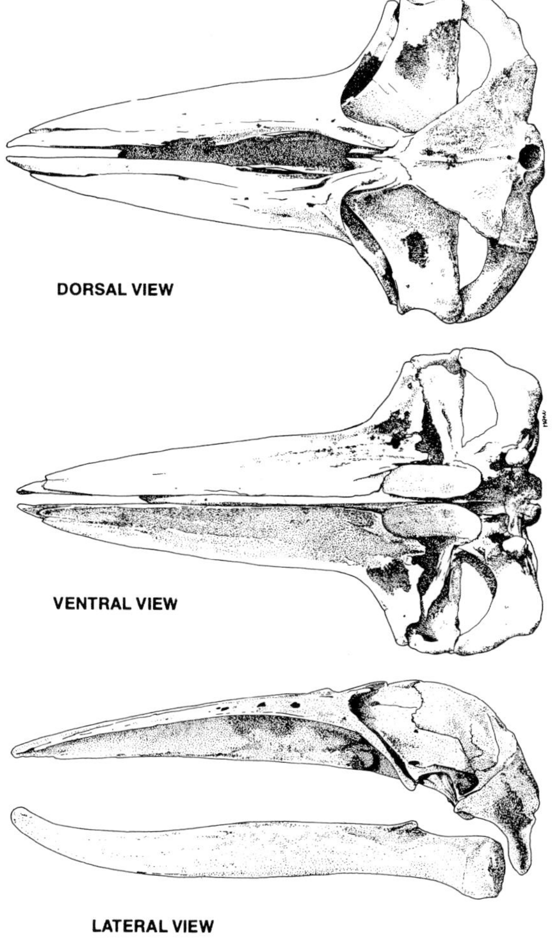


*3. Humpback whale,* Megaptera novaeangliae. *Drawn by M. D'Antoni,* FAO Species Identification Guide, Marine Mammals of the World, *Food and Agriculture Organization of the United Nations/United Nations Environment Programme, 1993*

b. Rostrum relatively flat; base of rostrum >½ cranial width; nasals reduced in size; frontals barely or not at all visible on vertex. Go to 3.

3a. Base of rostrum about ½ cranial width; anterior margin of squamosal rounded or U-shaped.

HUMPBACK WHALE, *Megaptera novaeangliae*

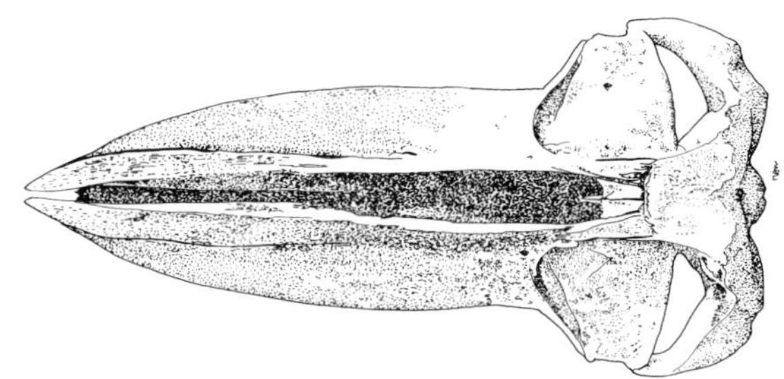

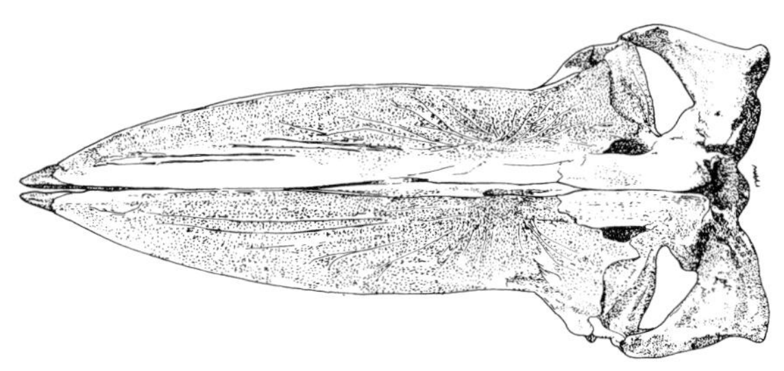


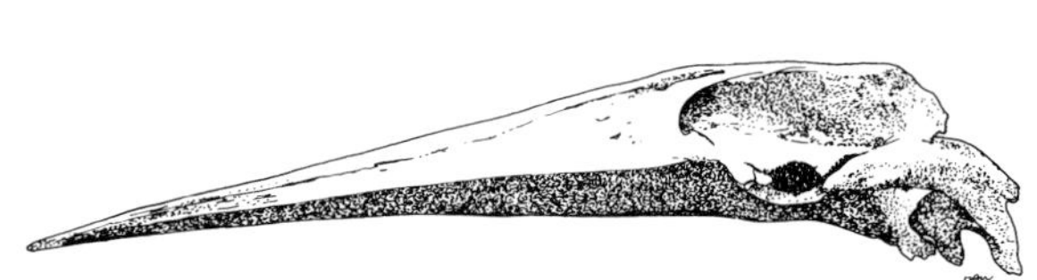


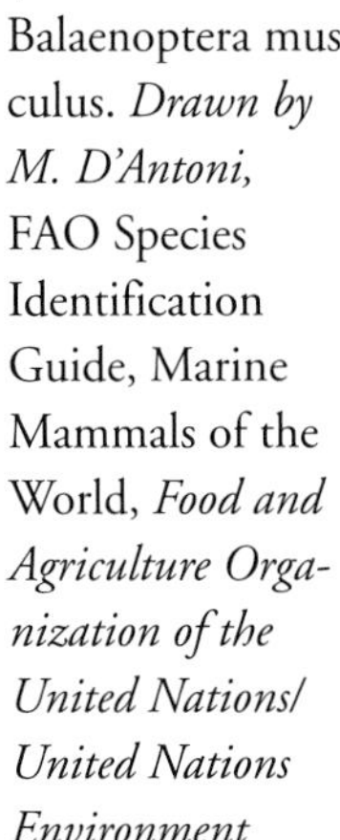

*4. Blue whale,* Balaenoptera musculus. *Drawn by M. D'Antoni,* FAO Species Identification Guide, Marine Mammals of the World, *Food and Agriculture Organization of the United Nations/ United Nations Environment Programme, 1993*

b. Base of rostrum >⅔ cranial width; anterior margin of squamosal pointed or V-shaped. Go to 4.

4a. Rostrum U-shaped and rounded at tip; rostral borders parallel along proximal half.

BLUE WHALE, *Balaenoptera musculus*

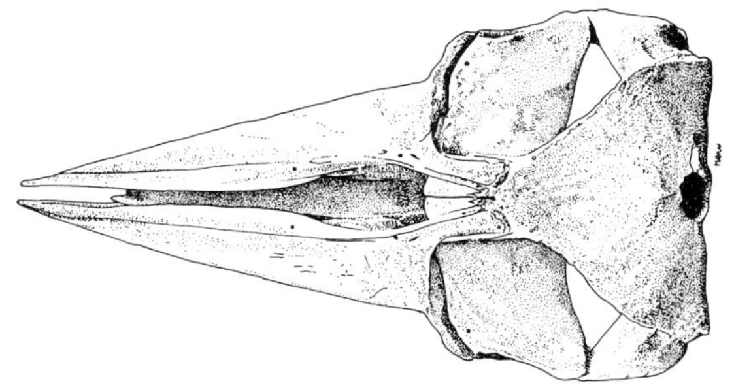


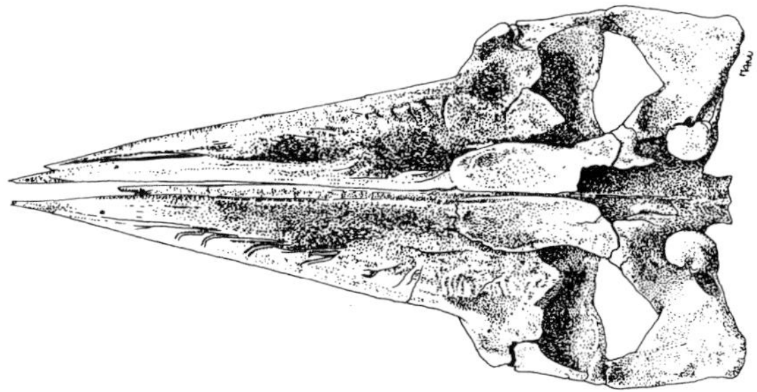


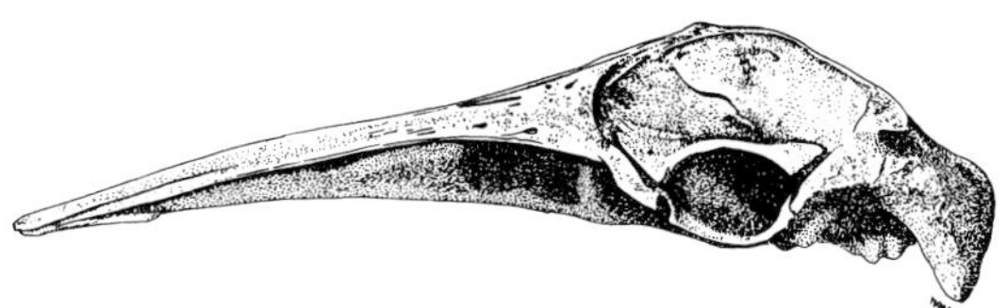


5. *Minke whale,* Balaenoptera acutorostrata. *Drawn by M. D'Antoni,* FAO Species Identification Guide, Marine Mammals of the World, *Food and Agriculture Organization of the United Nations/United Nations Environment Programme, 1993*

b. Rostrum V-shaped pointed; rostral borders divergent throughout length. Go to 5.

5a. Condylobasilar length (CBL) of adults <2 m.
MINKE WHALE, *Balaenoptera acutorostrata*

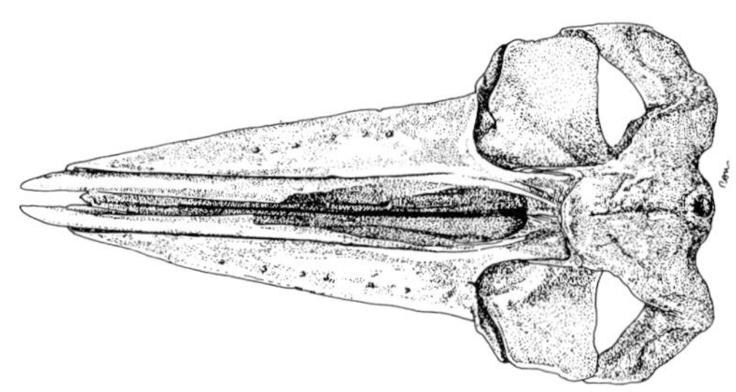


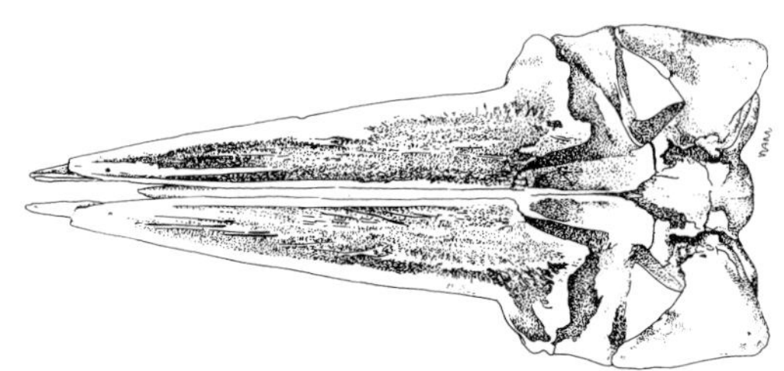


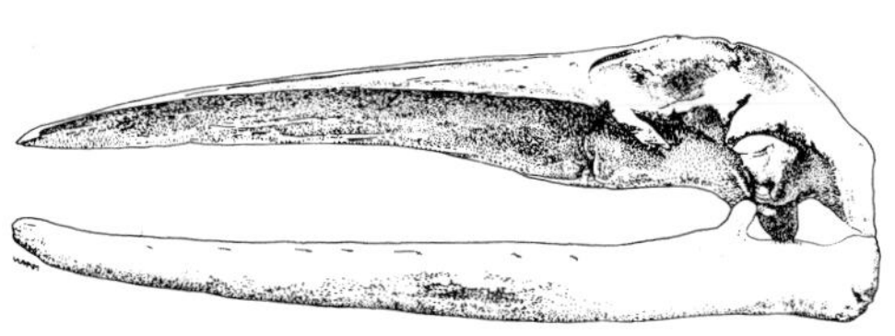


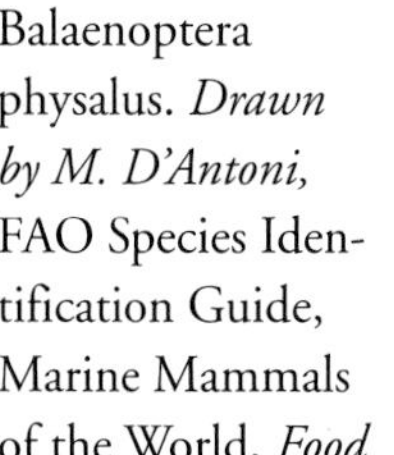

*6. Fin whale,* Balaenoptera physalus. *Drawn by M. D'Antoni,* FAO Species Identification Guide, Marine Mammals of the World, *Food and Agriculture Organization of the United Nations/United Nations Environment Programme, 1993*

b. CBL of adults >3 m.

FIN WHALE, *Balaenoptera physalus;* SEI WHALE, *Balaenoptera borealis* or; BRYDE'S WHALE, *Balaenoptera edeni*
(Note: It generally is not possible to reliably distinguish skulls of these species; consult an expert for identification.)

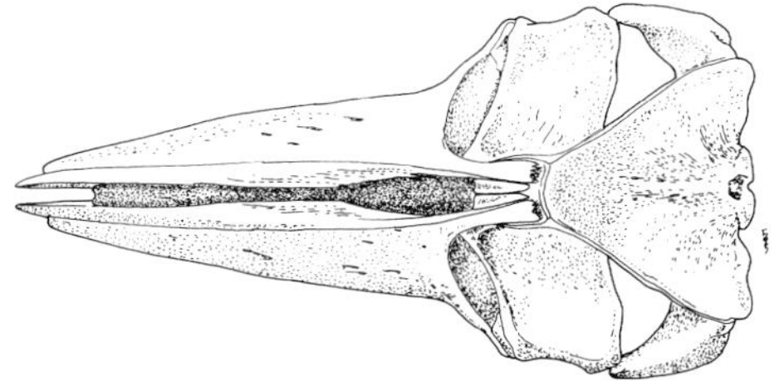


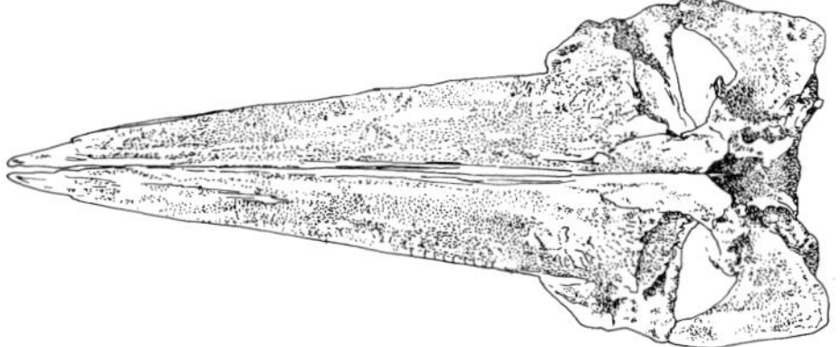


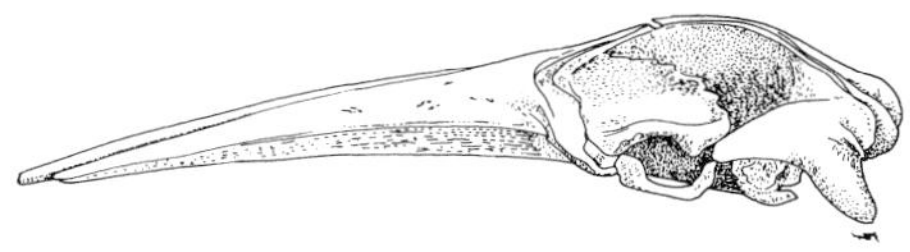


*7. Sei whale,* Balaenoptera borealis. *Drawn by M. D'Antoni,* FAO Species Identification Guide, Marine Mammals of the World, *Food and Agriculture Organization of the United Nations/ United Nations Environment Programme, 1993*

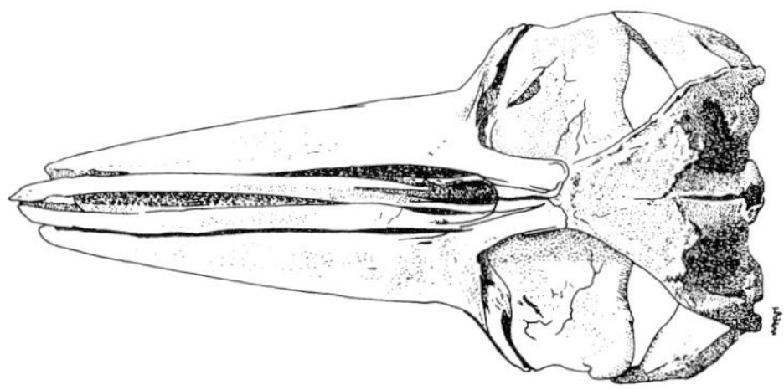


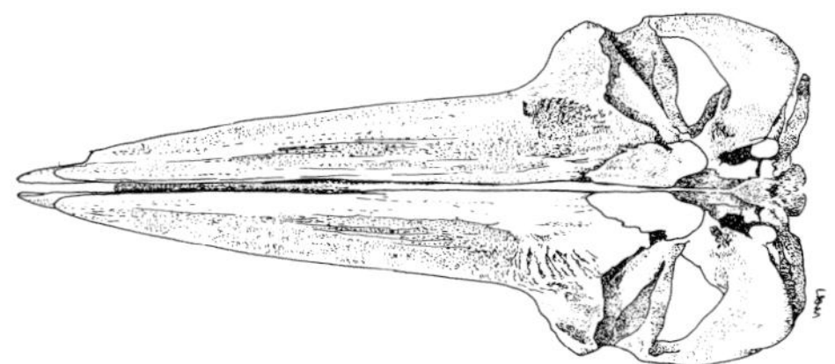


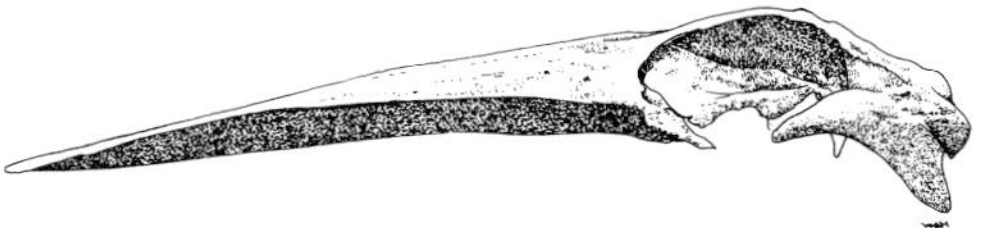


*8. Bryde's whale,* Balaenoptera edeni. *Drawn by M. D'Antoni,* FAO Species Identification Guide, Marine Mammals of the World, *Food and Agriculture Organization of the United Nations/ United Nations Environment Programme, 1993*

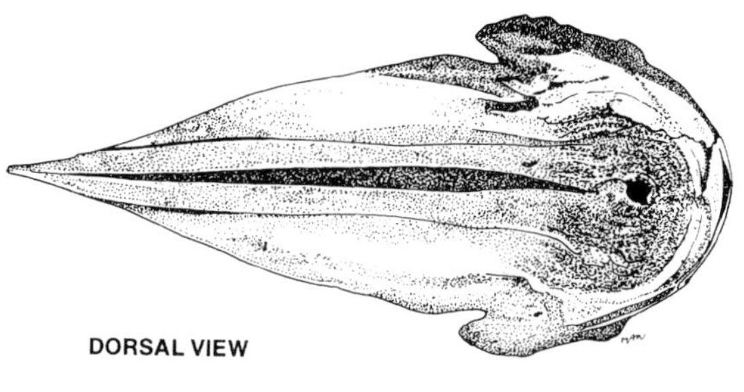


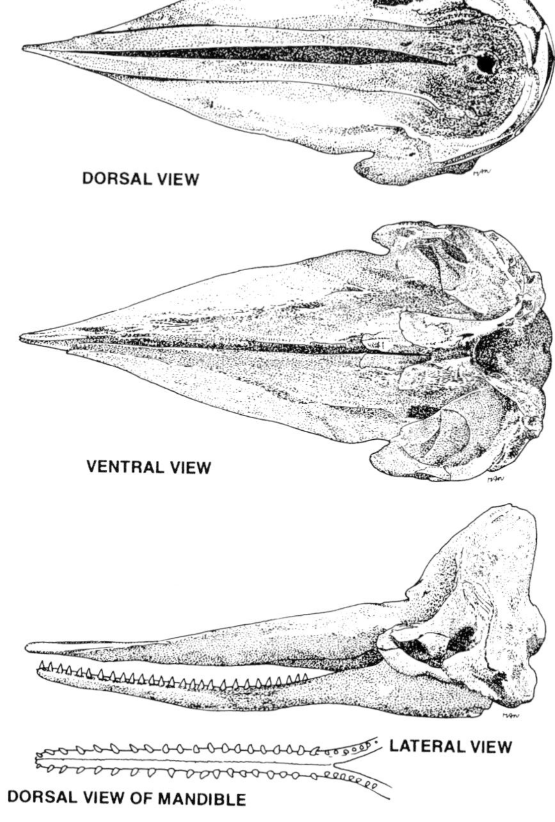


*9. Sperm whale,* Physeter macrocephalus. *Drawn by M. D'Antoni,* FAO Species Identification Guide, Marine Mammals of the World, *Food and Agriculture Organization of the United Nations/ United Nations Environment Programme, 1993*

6a. Anterior cranial region basinlike or with elevated maxillary ridges; rostrum deeper than wide in some species; teeth generally restricted to lower jaw. Go to 7.

b. Anterior cranial region neither basinlike nor with elevated maxillary ridges; rostrum wider than deep; teeth in both upper and lower jaws (except in *Grampus*). Go to 11.

7a. Nares extremely asymmetrical (left naris much larger than right); rostrum much wider than deep. Go to 8.

b. Nares similar in size; rostrum nearly as deep as wide. Go to 10.

8a. Rostrum long (>50 percent of CBL); zygomatic arches complete; >17 pairs of teeth; mandibular symphysis long (>30 percent of mandibular length).

SPERM WHALE, *Physeter macrocephalus*

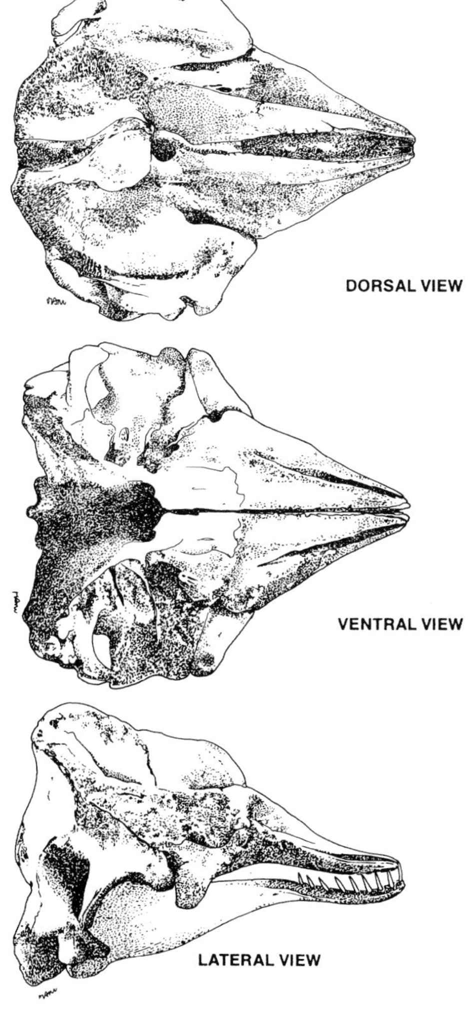


*10. Pygmy sperm whale,* Kogia breviceps. *Drawn by M. D'Antoni,* FAO Species Identification Guide, Marine Mammals of the World, *Food and Agriculture Organization of the United Nations/ United Nations Environment Programme, 1993*

b. Rostrum short (<50 percent of CBL); zygomatic arches incomplete; <17 pairs of teeth; mandibular symphysis short (<30 percent of mandibular length). Go to 9.

9a. Adult skull relatively large (CBL >39 cm); rostrum relatively long; typically 12–16 pairs of teeth (sometimes 10–11) only in lower jaw; teeth only slightly hooked.

PYGMY SPERM WHALE, *Kogia breviceps*

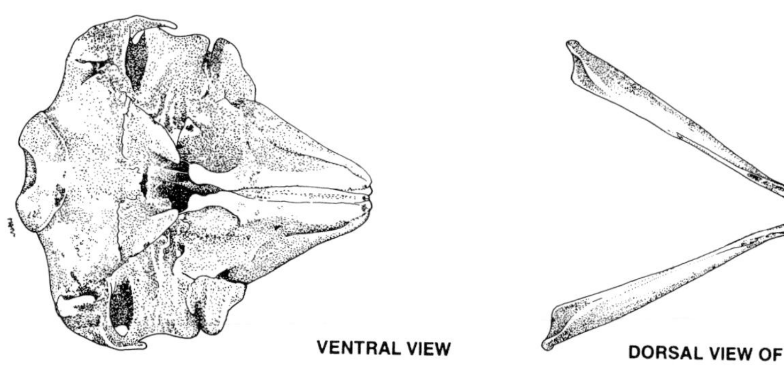


*11. Dwarf sperm whale,* Kogia simus. *Drawn by M. D'Antoni,* FAO Species Identification Guide, Marine Mammals of the World, *Food and Agriculture Organization of the United Nations/ United Nations Environment Programme, 1993*

b. Adult skull relatively small (CBL <31 cm); rostrum relatively short; typically 8–11 pairs of teeth (sometimes 12–13) in lower jaw and occasionally up to 3 vestigial pairs in upper jaw; teeth extremely hooked.

DWARF SPERM WHALE, *Kogia simus*

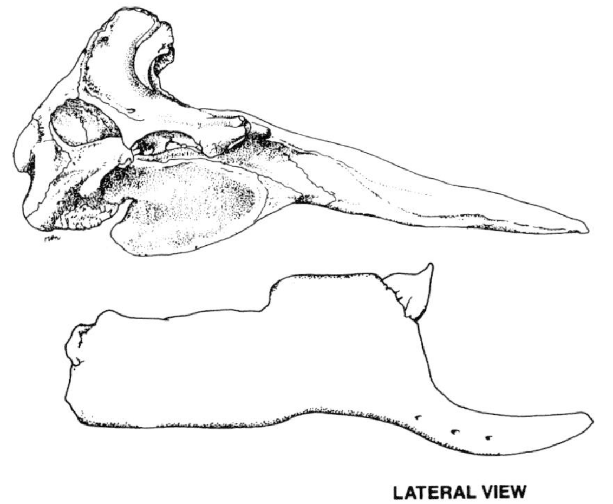
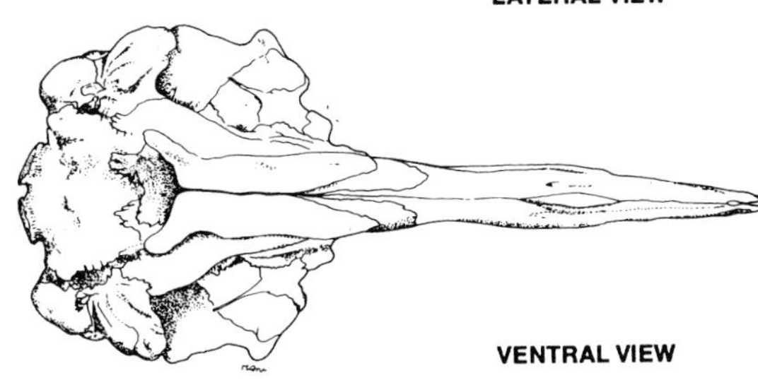


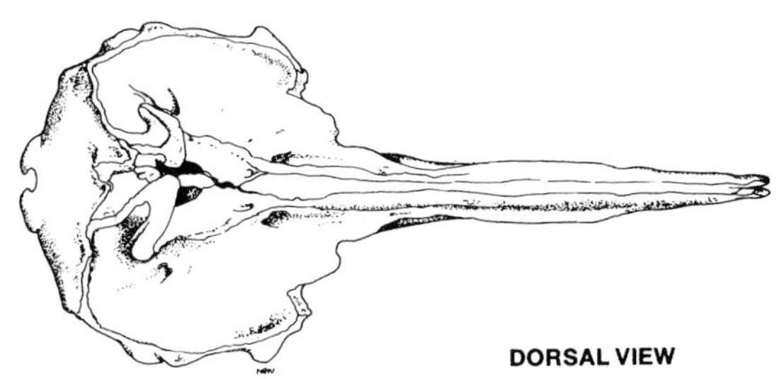


*12. Blainville's beaked whale,* Mesoplodon densirostris. *Drawn by M. D'Antoni,* FAO Species Identification Guide, Marine Mammals of the World, *Food and Agriculture Organization of the United Nations/United Nations Environment Programme, 1993*

10a. Teeth (1 pair, if present) flattened and located well back from tip of mandibles.
*Mesoplodon* sp. (Note: It generally is not possible to reliably distinguish skulls of these species; consult an expert for identification.)

I. Wide tusks of males located on massive bony arches posterior to symphysis, and lean forward.
BLAINVILLE'S BEAKED WHALE, *Mesoplodon densirostris*

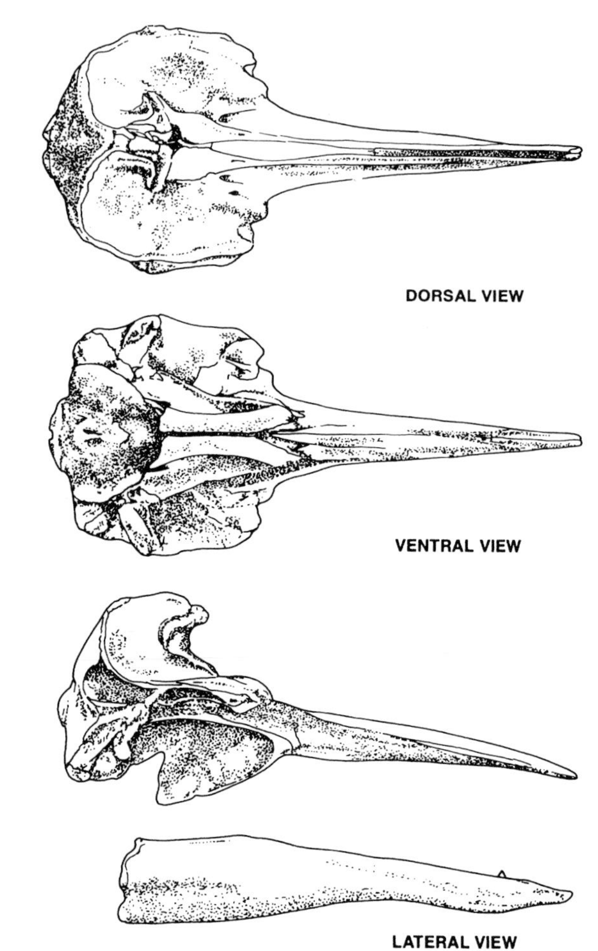


*13. Gervais' beaked whale,* Mesoplodon europeaus. *Drawn by M. D'Antoni,* FAO Species Identification Guide, Marine Mammals of the World, *Food and Agriculture Organization of the United Nations/ United Nations Environment Programme, 1993*

II. Flattened tusks of males between anterior end and symphysis.
GERVAIS' BEAKED WHALE, *Mesoplodon europaeus*

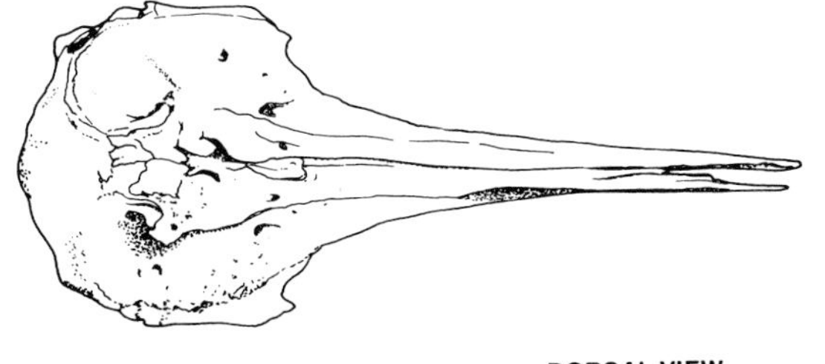


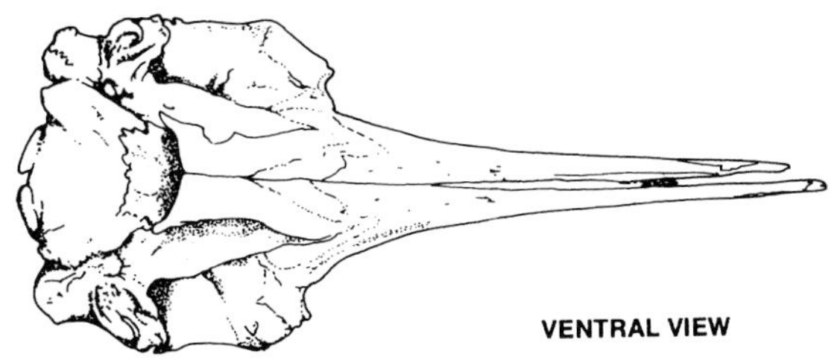


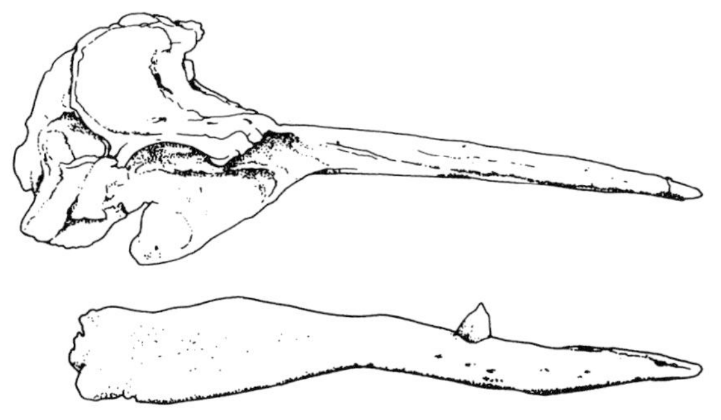


*14. Sowerby's beaked whale,* Mesoplodon bidens. *Drawn by M. D'Antoni,* FAO Species Identification Guide, Marine Mammals of the World, *Food and Agriculture Organization of the United Nations/ United Nations Environment Programme, 1993*

III. Flattened tusks of males overlap symphysis.
SOWERBY'S BEAKED WHALE, *Mesoplodon bidens*

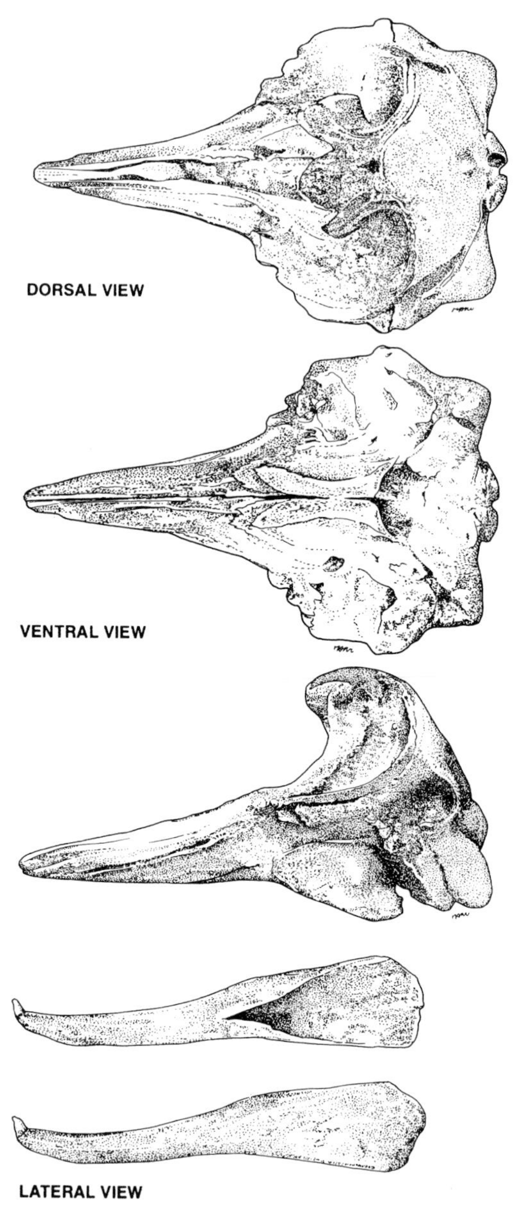


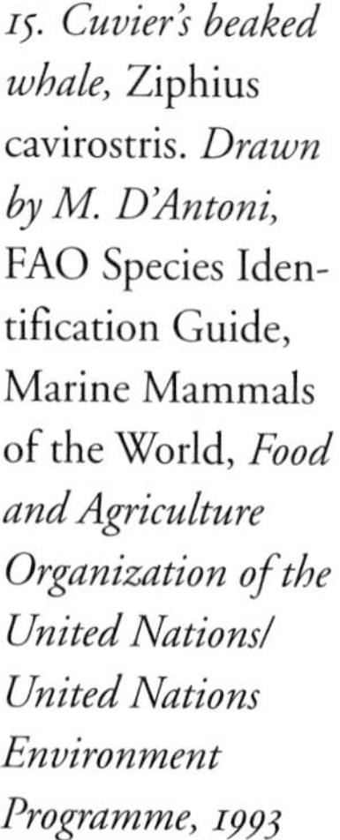
*15. Cuvier's beaked whale,* Ziphius cavirostris. *Drawn by M. D'Antoni,* FAO Species Identification Guide, Marine Mammals of the World, *Food and Agriculture Organization of the United Nations/ United Nations Environment Programme, 1993*

b. Teeth (1–2 pairs) oval to round in cross section at tip of mandibles.

CUVIER'S BEAKED WHALE, *Ziphius cavirostris*

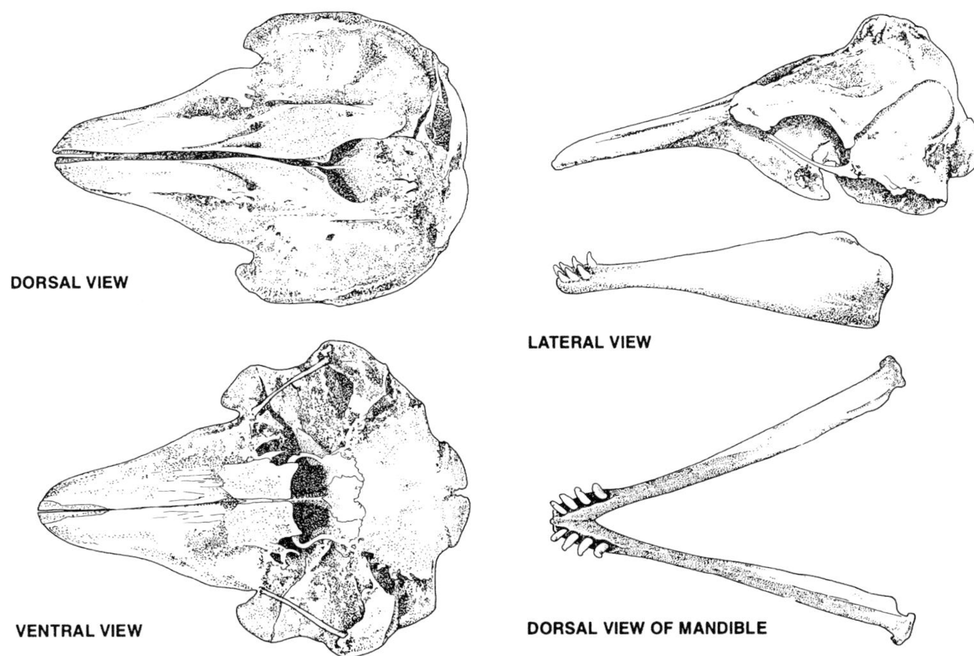


*16. Risso's dolpin,* Grampus griseus. *Drawn by M. D'Antoni,* FAO Species Identification Guide, Marine Mammals of the World, *Food and Agriculture Organization of the United Nations/United Nations Environment Programme, 1993*

11a. Teeth/row <29. Go to 12.

b. Teeth/row >29. Go to 19.

12a. Teeth (2–7 pairs) near tip of lower jaw only (uncommonly 1–2 pairs in upper jaw); lateral margins of rostrum concave along middle part of their length.

RISSO'S DOLPHIN, *Grampus griseus*

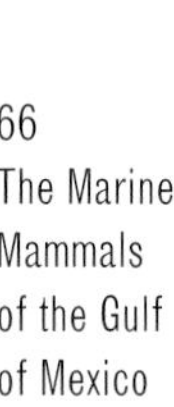


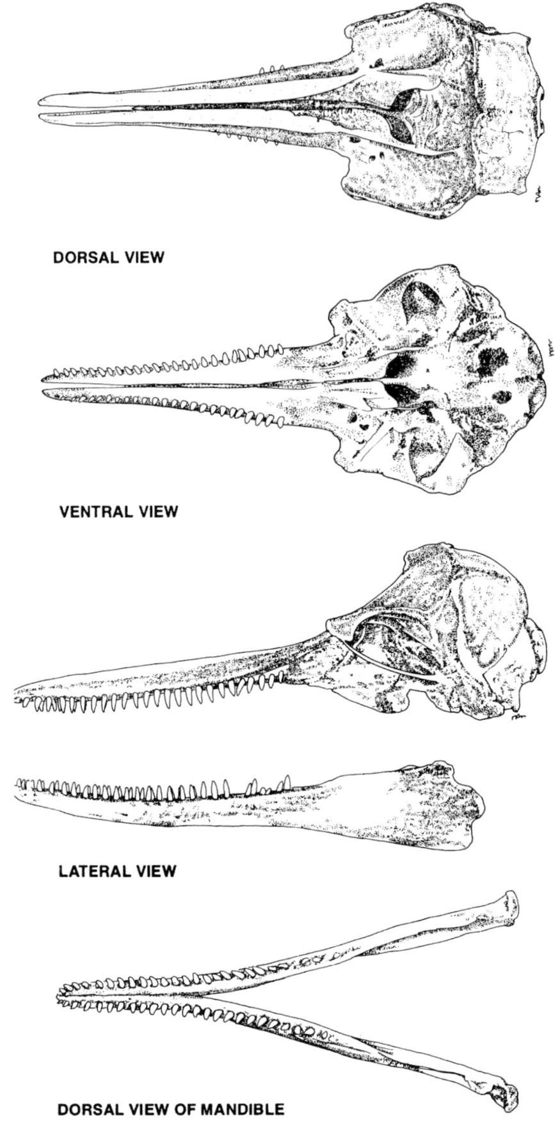


*17. Rough-toothed dolphin,* Steno bredanensis. *Drawn by M. D'Antoni,* FAO Species Identification Guide, Marine Mammals of the World, *Food and Agriculture Organization of the United Nations/ United Nations Environment Programme, 1993*

b. Teeth/row (>7) on both upper and lower jaws; lateral margins of rostrum generally convex. Go to 13.

13a. Teeth with shallow vertical wrinkles; mandibular symphysis long (>25 percent of mandibular length); 19–28 teeth/row; constriction at base of rostrum.

ROUGH-TOOTHED DOLPHIN, *Steno bredanensis*

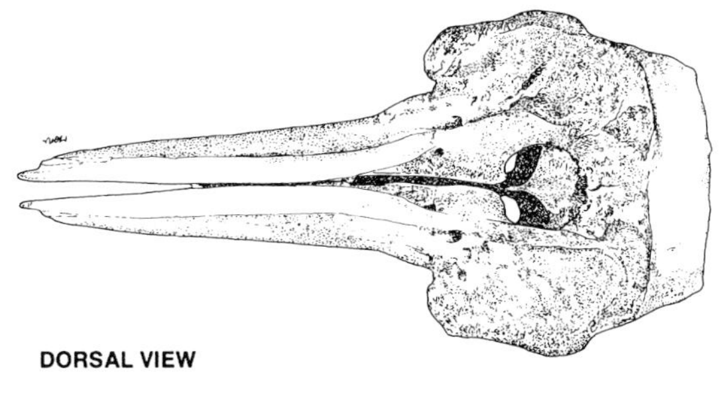


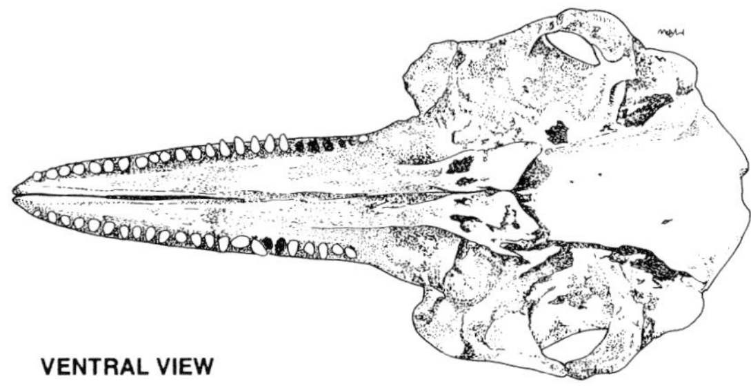


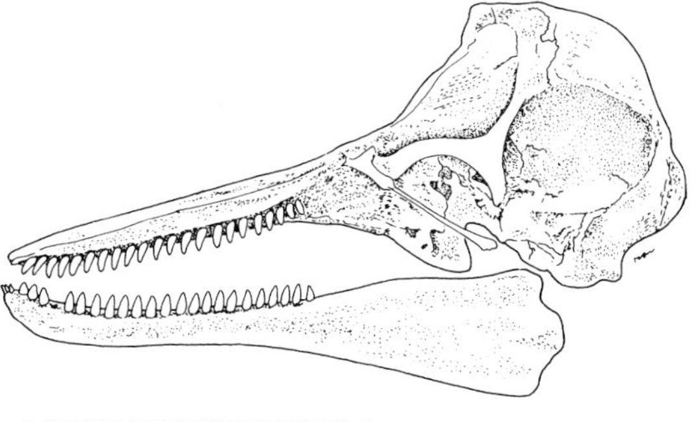


*18. Bottlenose dolphin,* Tursiops truncatus. *Drawn by M. D'Antoni,* FAO Species Identification Guide, Marine Mammals of the World, *Food and Agriculture Organization of the United Nations/ United Nations Environment Programme, 1993*

b. Teeth without wrinkles; mandibular symphysis relatively short (<25 percent of mandibular length); no constriction at base of rostrum. Go to 14.

14a. Teeth/row >15. Go to 15.

b. Teeth/row <15. Go to 16.

15a. Rostrum relatively narrow (length/breadth ratio >2:1); 18–26 teeth/row; antorbital notches relatively shallow.
BOTTLENOSE DOLPHIN, *Tursiops truncatus*

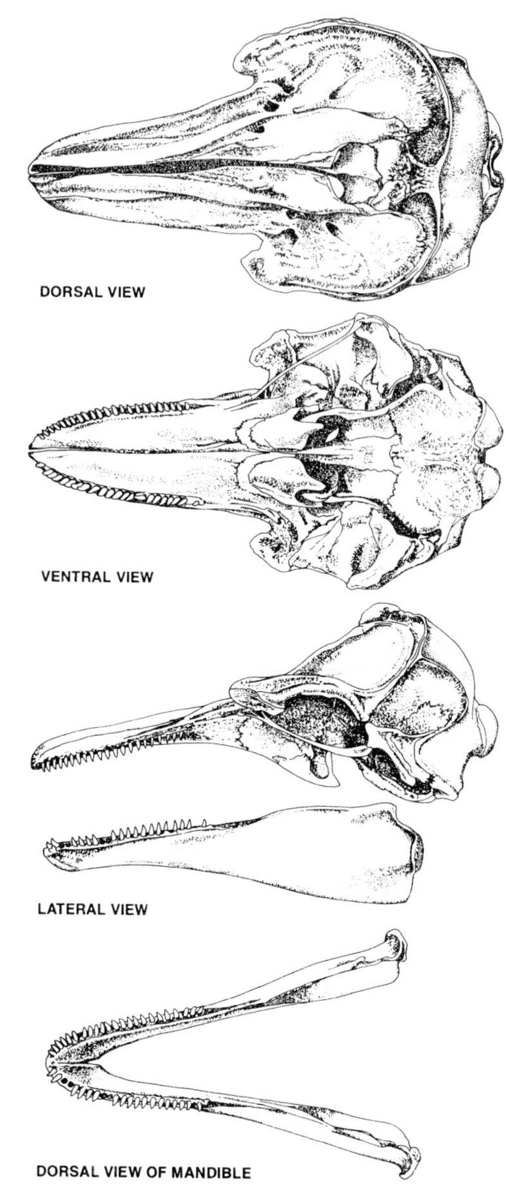


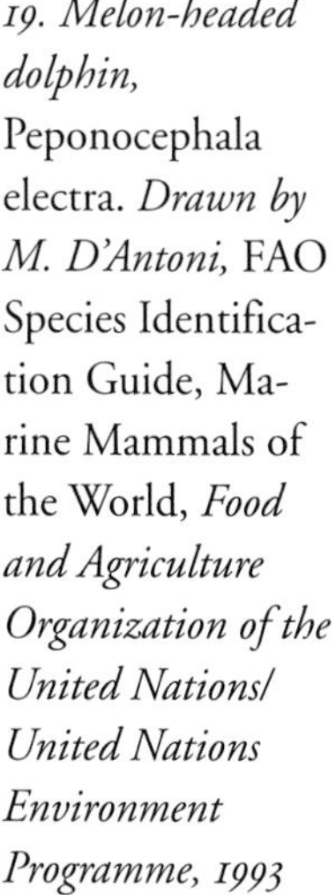
*19. Melon-headed dolphin,* Peponocephala electra. *Drawn by M. D'Antoni,* FAO Species Identification Guide, Marine Mammals of the World, *Food and Agriculture Organization of the United Nations/ United Nations Environment Programme, 1993*

b. Rostrum relatively wide (length/breadth ratio <2:1); 20–26 teeth/row; antorbital notches very deep.
MELON-HEADED WHALE, *Peponocephala electra*

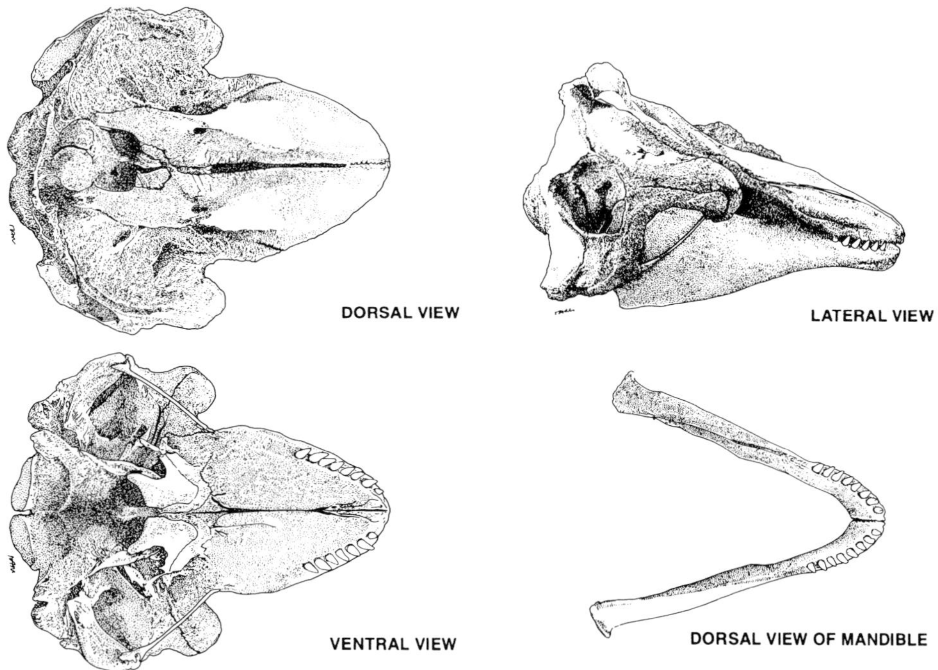


*20. Short-finned pilot whale,* Globicephala macrorhynchus. *Drawn by M. D'Antoni,* FAO Species Identification Guide, Marine Mammals of the World, *Food and Agriculture Organization of the United Nations/ United Nations Environment Programme, 1993*

16a. Teeth (7–13) only in anterior half of rostrum. Go to 17.

b. Teeth in both halves of rostrum; rostrum relatively narrow (length/breadth ratio generally >1.3:1); from dorsal view, maxillae largely exposed along distal half of rostrum. Go to 18.

17a. Teeth (7–9) only in anterior half of rostrum; skull very wide (length/breadth ratio generally ca. 1:1); from dorsal view, premaxillae nearly cover maxillae along distal half of rostrum; from lateral view, lower jaw high and short (length/height ratio in dorsoventral plane ca. 3:1).

SHORT-FINNED PILOT WHALE, *Globicephala macrorhynchus*

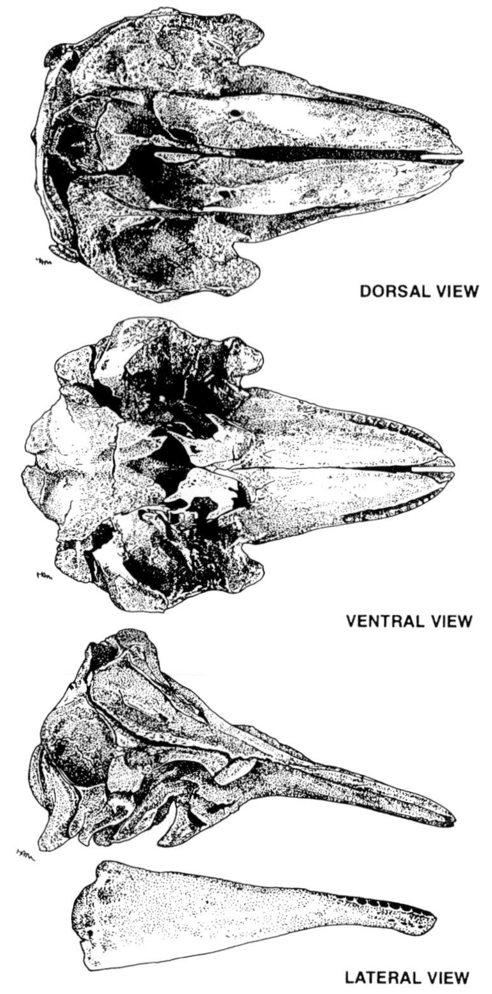


*21. Long-finned pilot whale,* Globicephala melas. *Drawn by M. D'Antoni,* FAO Species Identification Guide, Marine Mammals of the World, *Food and Agriculture Organization of the United Nations/United Nations Environment Programme, 1993*

b. Teeth (8–13) only in anterior half of rostrum; skull length to breadth ratio close to 1.5:1; from lateral view, lower jaw length/height ratio about 3.6:1.

LONG-FINNED PILOT WHALE, *Globicephala melas*

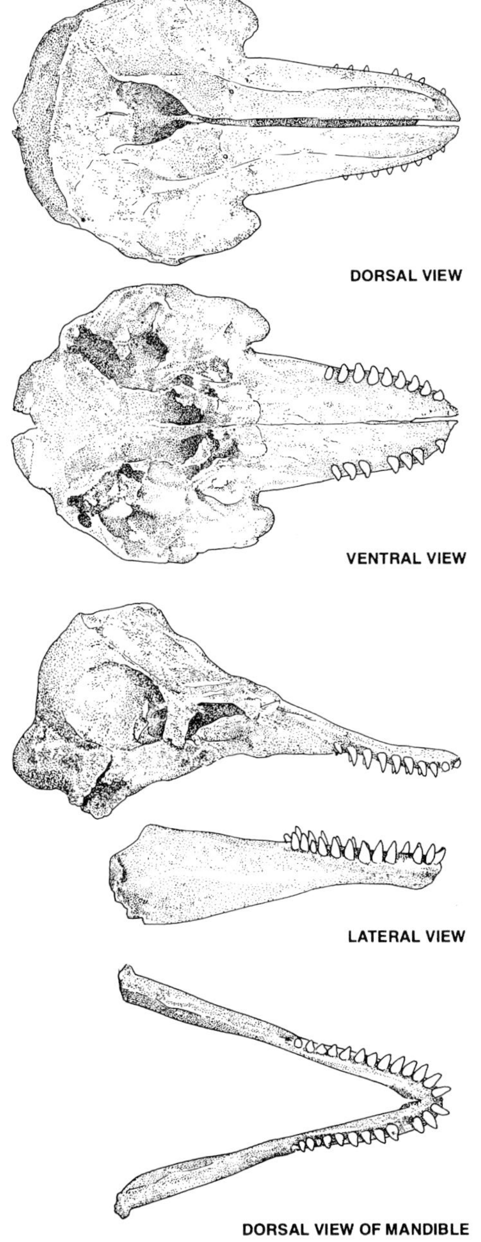


*22. Pygmy killer whale,* Feresa attenuata. *Drawn by M. D'Antoni,* FAO Species Identification Guide, Marine Mammals of the World, *Food and Agriculture Organization of the United Nations/United Nations Environment Programme, 1993*

18a. Teeth relatively slender (generally <10 mm in diameter); adult CBL <50 cm; 8–13 teeth in anterior ⅔ of rostral tooth rows only.

PYGMY KILLER WHALE, *Feresa attenuata*

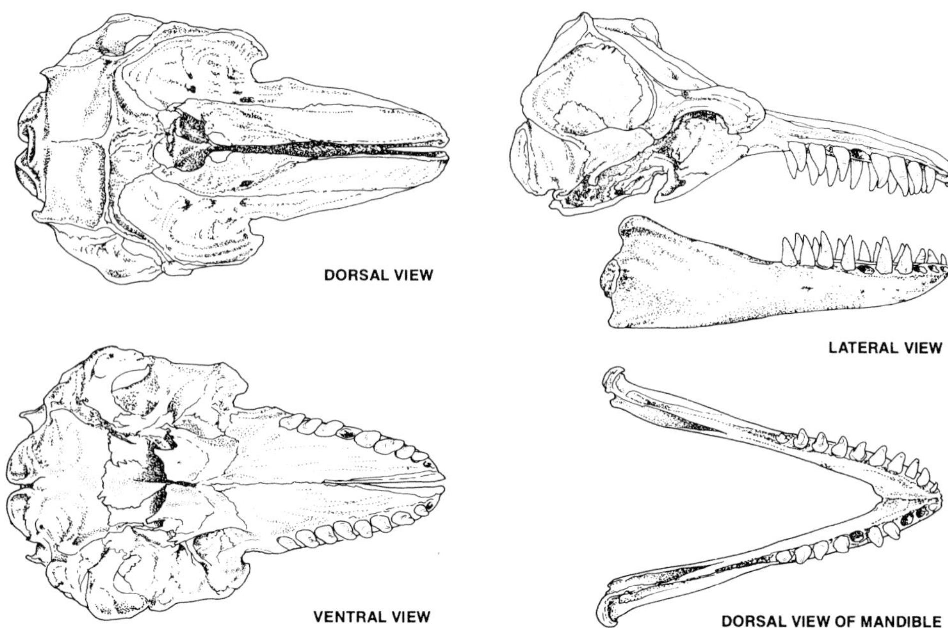


*23. False killer whale,* Pseudorca crassidens. *Drawn by M. D'Antoni,* FAO Species Identification Guide, Marine Mammals of the World, *Food and Agriculture Organization of the United Nations/ United Nations Environment Programme, 1993*

b. Teeth relatively robust (generally >10 mm in diameter); adult CBL >50 cm; teeth present in posterior ⅔ of rostral tooth rows. Go to 19.

19a. Teeth round in cross section (greatest diameter of largest teeth generally <23 mm); adult CBL <78 cm; 7–12 teeth/row; width across premaxillae >50 percent of rostral basal width.
FALSE KILLER WHALE, *Pseudorca crassidens*

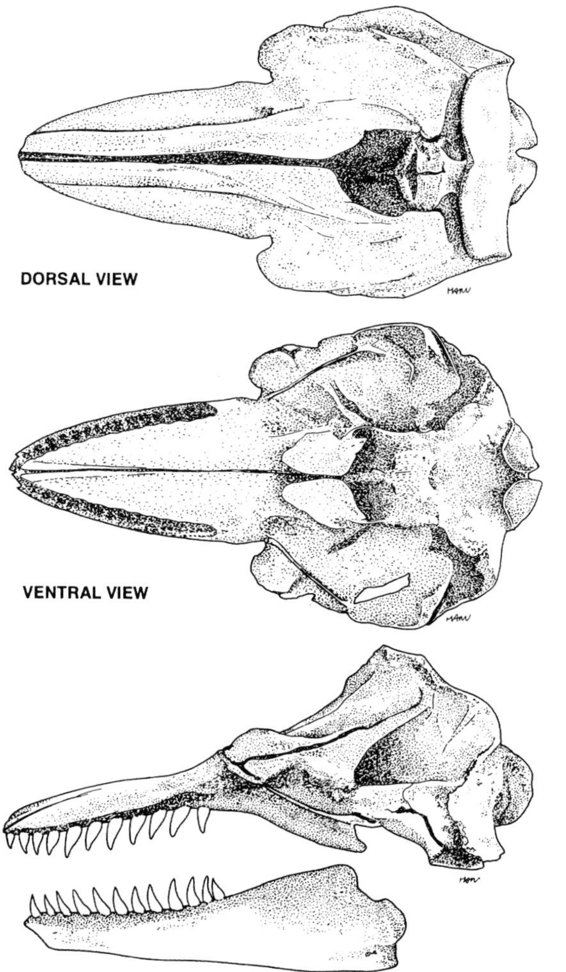


*24. Killer whale,* Orcinus orca. *Drawn by M. D'Antoni,* FAO Species Identification Guide, Marine Mammals of the World, *Food and Agriculture Organization of the United Nations/United Nations Environment Programme, 1993*

b. Teeth oval in cross section (greatest diameter of largest teeth generally >23 mm); adult CBL >78 cm; 10–14 teeth/row; width across premaxillae <50 percent of rostral basal width.
KILLER WHALE, *Orcinus orca*

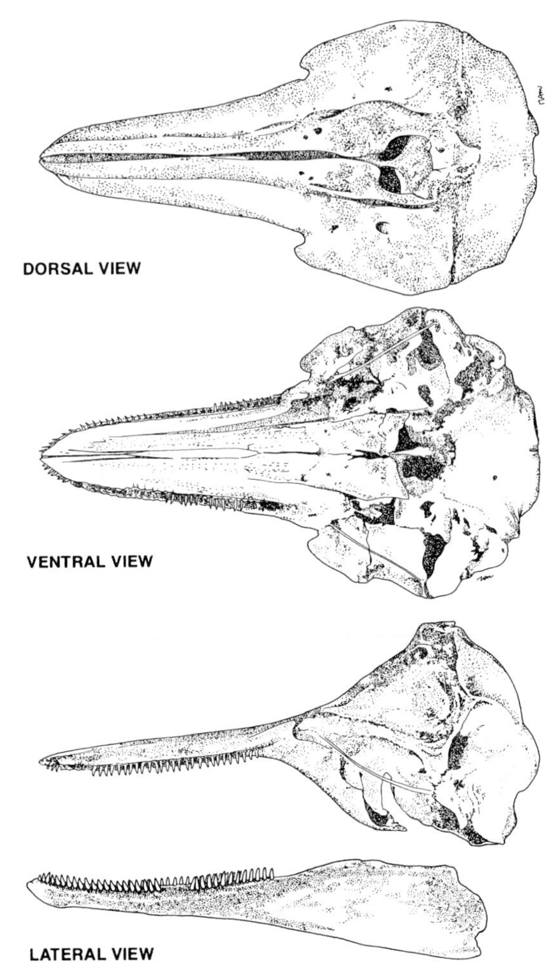


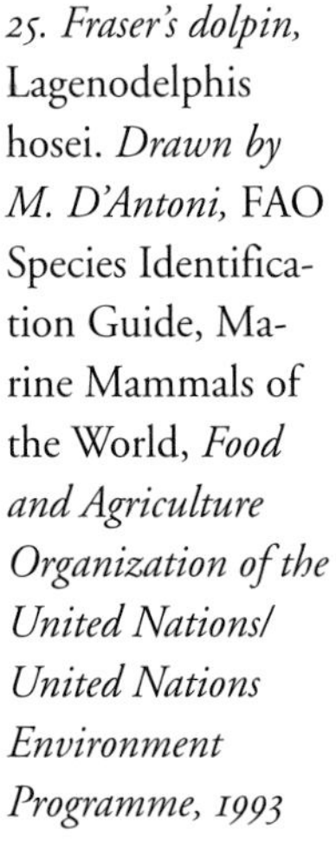

*25. Fraser's dolpin,* Lagenodelphis hosei. *Drawn by M. D'Antoni,* FAO Species Identification Guide, Marine Mammals of the World, *Food and Agriculture Organization of the United Nations/ United Nations Environment Programme, 1993*

20a. Palatal grooves deep (>3 mm at middle of rostrum). Go to 21.

b. Palatal grooves shallow (<3 mm at middle of rostrum) or nonexistent. Go to 23.

21a. Rostrum relatively wide (length/breadth ratio <2.4:1); <45 teeth/row.

FRASER'S DOLPHIN, *Lagenodelphis hosei*

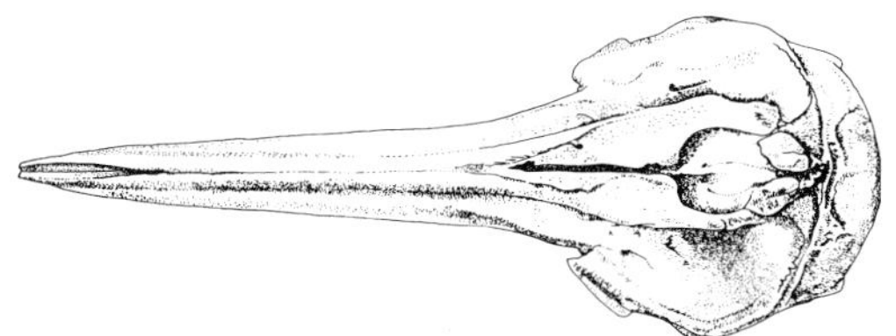


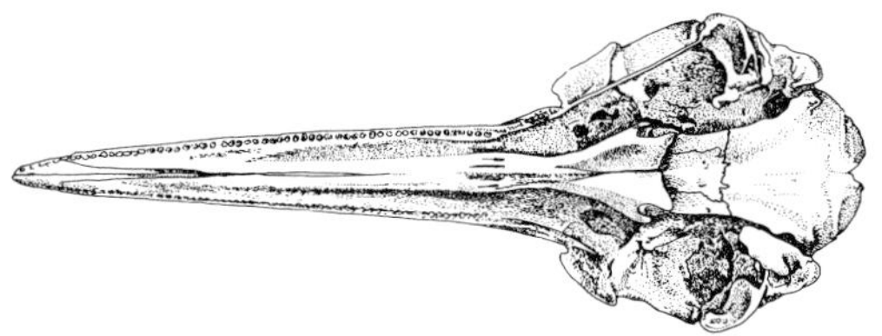


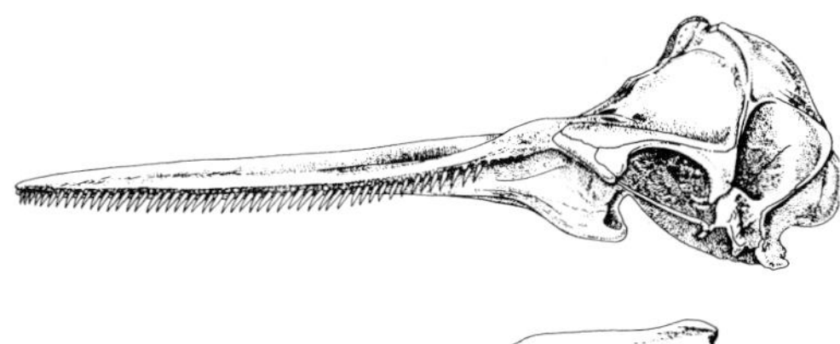

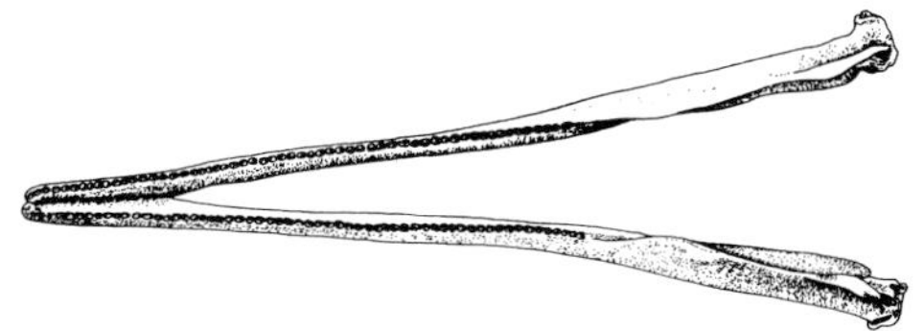


*26. Long-beaked common dolphin,* Delphinus capensis. *Drawn by M. D'Antoni,* FAO Species Identification Guide, Marine Mammals of the World, *Food and Agriculture Organization of the United Nations/ United Nations Environment Programme, 1993*

b. Rostrum relatively narrow (length/breadth ratio >2.5:1); >40 teeth/row. Go to 22.

22a. Rostrum relatively short and wide (<275 mm; length/breadth ratio <3.2:1); 41–54 teeth/row.
SHORT-BEAKED COMMON DOLPHIN, *Delphinus delphis*

b. Rostrum relatively long and slender (>275 mm; length/breadth ratio >3.2:1); 47–60 teeth/row.
LONG-BEAKED COMMON DOLPHIN, *Delphinus capensis*

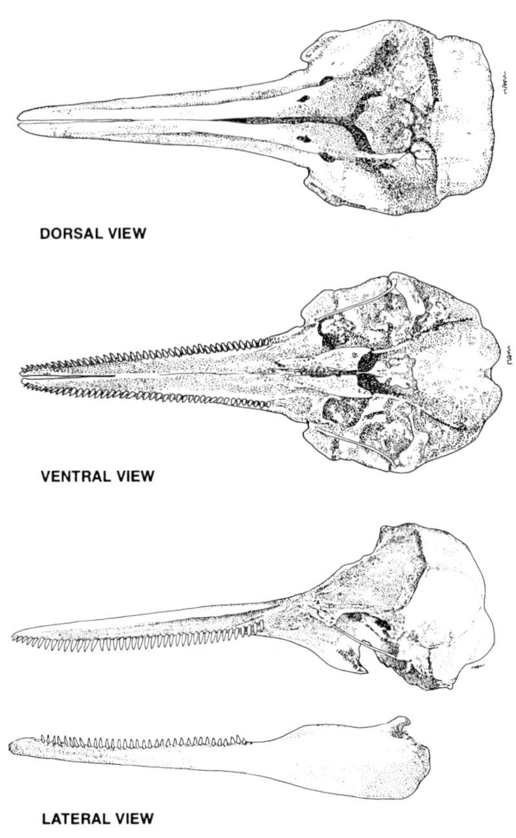


*27. Pantropical spotted dolphin,* Stenella attenuata. *Drawn by M. D'Antoni,* FAO Species Identification Guide, Marine Mammals of the World, *Food and Agriculture Organization of the United Nations/ United Nations Environment Programme, 1993*

23a. Mandibular symphysis relatively long (usually >17 percent of mandible length); mandible arcuate; <49 teeth/row; temporal fossae relatively large (length >14 percent CBL); distal half of rostrum rounded on dorsal surface; no palatal grooves; tooth rows converge throughout their length. Go to 24.

b. Mandibular symphysis relatively short (usually <17 percent of mandible length); mandible sigmoid; >35 teeth/row (usually >40); temporal fossae relatively small (length <17 percent CBL); distal half of rostrum typically flattened on dorsal surface; palatal grooves sometimes present; central portions of tooth rows parallel or nearly parallel. Go to 25.

24a. Teeth/row 34–48; teeth relatively small (2.6–4.1 mm in diameter); rostrum narrow distally (width at ¾ length <16 percent length).

PANTROPICAL SPOTTED DOLPHIN, *Stenella attenuata*

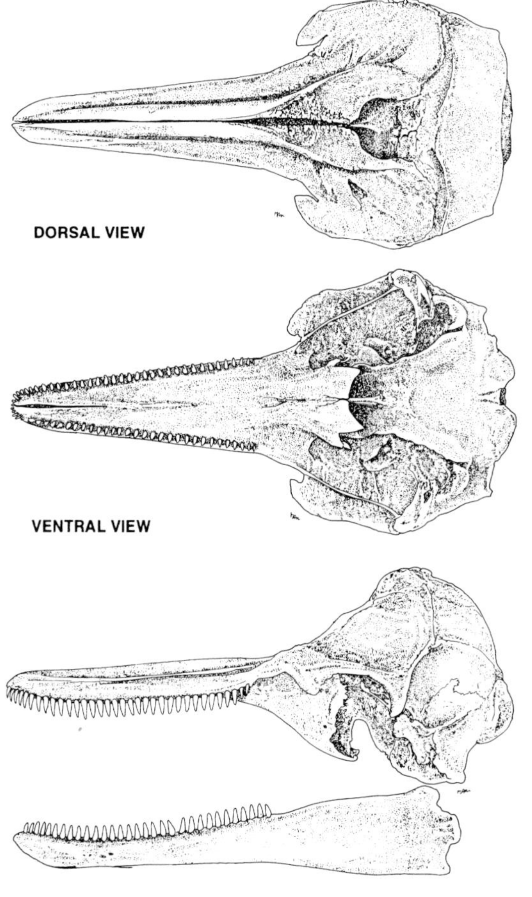


*28. Atlantic spotted dolphin,* Stenella frontalis. *Drawn by M. D'Antoni,* FAO Species Identification Guide, Marine Mammals of the World, *Food and Agriculture Organization of the United Nations/ United Nations Environment Programme, 1993*

b. Teeth/row 30–42; teeth relatively large (3.2–5.3 mm in diameter); rostrum broad distally (width at ¾ length >14 percent length).

ATLANTIC SPOTTED DOLPHIN, *Stenella frontalis*

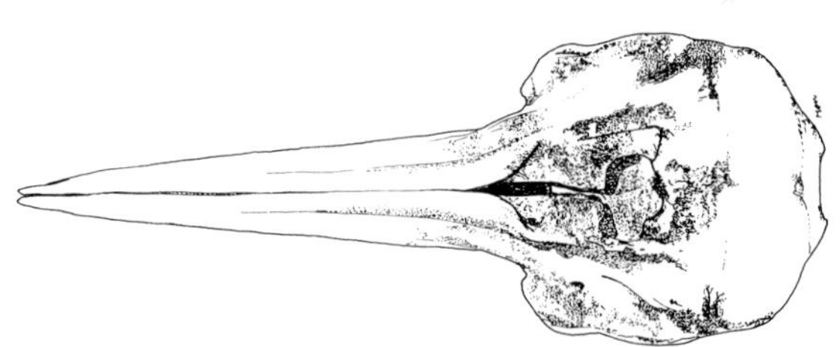


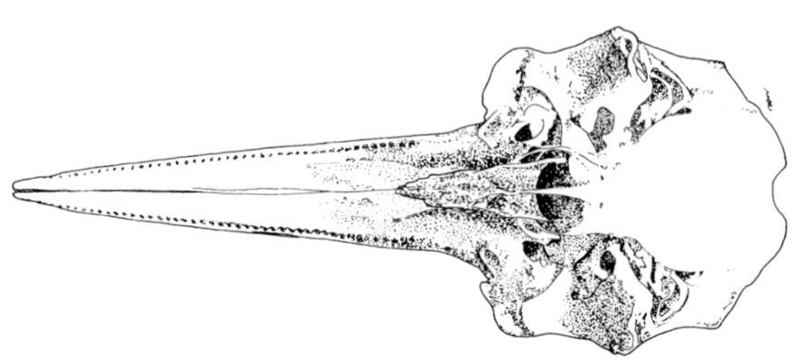


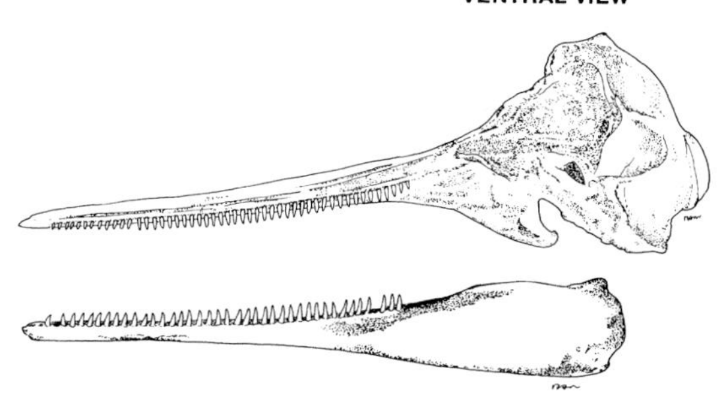


*29. Spinner dolphin,* Stenella longirostris. *Drawn by M. D'Antoni,* FAO Species Identification Guide, Marine Mammals of the World, *Food and Agriculture Organization of the United Nations/ United Nations Environment Programme, 1993*

25a. Rostrum relatively long and slender (>61 percent CBL; length/breadth ratio >3:1); 42–64 teeth/row; preorbital width <158 mm; upper tooth row >215 mm.

SPINNER DOLPHIN, *Stenella longirostris*

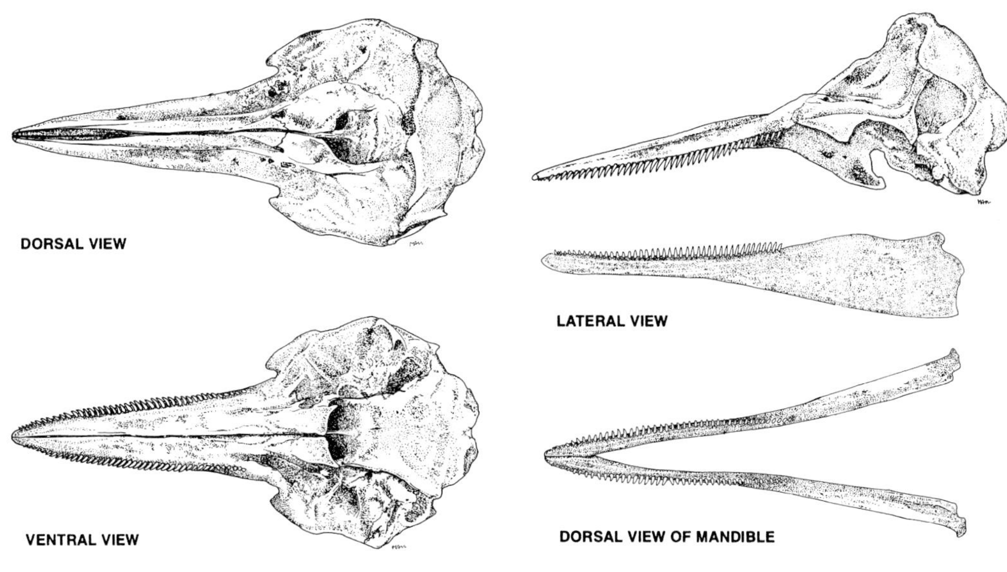


*30. Striped dolphin,* Stenella coeruleoalba. *Drawn by M. D'Antoni,* FAO Species Identification Guide, Marine Mammals of the World, *Food and Agriculture Organization of the United Nations/United Nations Environment Programme, 1993*

b. Rostrum relatively short and wide (<62 percent CBL; length/breadth ratio <3:1); <56 teeth/row; preorbital width >149 mm. Go to 26.

26a. Adult skull relatively large (CBL >420 mm); 39–55 teeth/row; palatal grooves usually shallow (<0.5 mm at ½ length of rostrum); upper tooth row >212 mm; preorbital width >177 mm; often a raised boss on premaxillae near base of rostrum (visible in lateral view).

STRIPED DOLPHIN, *Stenella coeruleoalba*

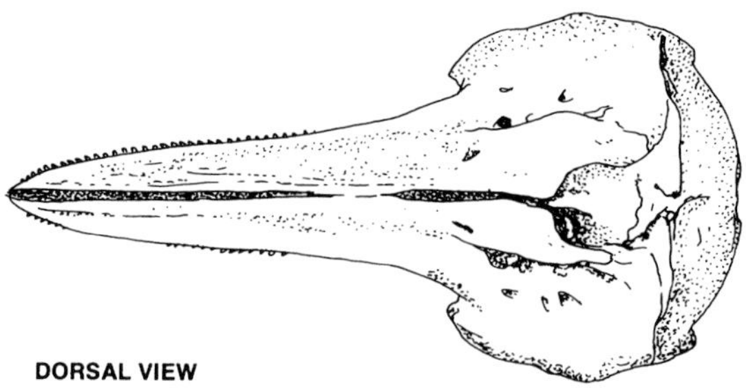


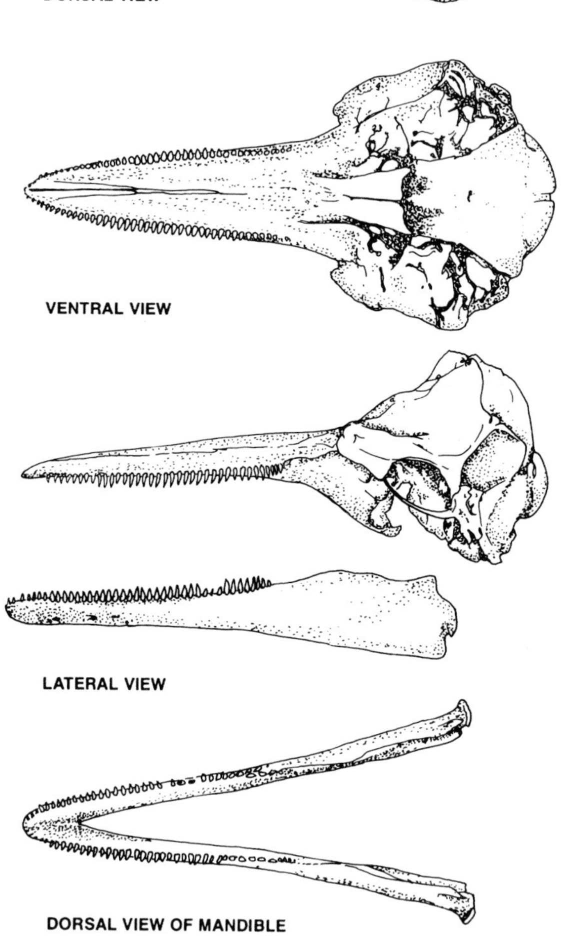


*31. Clymene dolphin,* Stenella clymene. *Drawn by M. D'Antoni,* FAO Species Identification Guide, Marine Mammals of the World, *Food and Agriculture Organization of the United Nations/United Nations Environment Programme, 1993*

b. Adult skull relatively small (CBL <415 mm); 36–52 teeth/row; palatal grooves usually distinct and relatively deep (>0.5 mm at ½ length of rostrum); upper tooth row >212 mm; preorbital width <175 mm; rostrum relatively flat when viewed from lateral aspect.

CLYMENE DOLPHIN, *Stenella clymene*

# Chapter 4

# History of Cetacean Research in the Gulf of Mexico

*"There's the story of the Irishman who was asked whether there were any leprechauns in Ireland. To which the canny Gael replied, 'There are no such things as leprechauns, but we know they are there anyway.' The same goes for whales in the Gulf of Mexico."*

*—Ednard Waldo (1957)*
*Whales in the Gulf of Mexico*

OF THE EAST, WEST, AND GULF COASTS of the United States, the cetacean fauna of the Gulf of Mexico has been by far the most poorly known, until recently. The only species that could be said to be well studied in the Gulf of Mexico is a coastal species, the bottlenose dolphin (*Tursiops truncatus*). The other twenty-seven species of cetaceans documented in the Gulf of Mexico occur primarily offshore in relatively deep waters.

Prior to 1977, there was no systematic effort to document cetacean records in the Gulf of Mexico. The only information came from opportunistic reporting of strandings, sightings, and occasional captures. Because of the sparsity of marine mammal scientists along the shores of the Gulf and the remoteness of much of its coastline, the literature is not as extensive as for other coasts. It is largely through the dedication of a small number of individuals with an interest in marine mammals that we have as much to go on as we do. Researchers such as D. K. and M. C. Caldwell, J. C.

*32. Nearshore bottlenose dolphin mother and calf leaping from the water. This kind of exuberant play is a common sight near shore just about anywhere along the Gulf of Mexico. Courtesy Bernd Würsig*

Moore, and J. N. Layne in Florida, G. H. Lowery in Louisiana, G. Gunter in Texas (and later Mississippi), and D. J. Schmidly in Texas (all in the bibliography) accumulated records and published summaries for certain subsections of the Gulf. The Mexican coastline has remained almost completely unstudied, and even less is known about the Cuban Gulf Coast.

Essentially, all of these data were collected opportunistically, and in most cases tell us little more than the cetacean species composition of the Gulf. For some species, enough records have accumulated to reveal something of long-term population trends, seasonal occurrence, or relative abundance within the Gulf.

Since 1977, with the establishment of the Southeastern United States Marine Mammal Stranding Network (SEUS MMSN), systematic stranding data have been collected and archived. Schmidly in 1981 summarized information available through 1979 on cetacean distribution in the Gulf of Mexico based largely on strandings, and Jefferson and colleagues updated Schmidly's maps in 1992. Stranding data, however, are subject to potential biases; stranded animals are usually sick or injured, and at times far from their normal habitat. These data provided an indication of the species composition of the Gulf and seasonal and geographic differences for some species, but little on the true habitat preferences or abundances of different cetacean species.

Many localized and some synoptic studies have described the coastal waters of the Gulf of Mexico, mainly for manatees in Florida and bottlenose dolphins throughout the Gulf. Only a few systematic sighting surveys, however, have been conducted in offshore Gulf waters, and the first of these started as recently as the late 1970s. These are described below, with locations of survey effort depicted in Figure 3.



## USFWS Aerial Surveys ("The Fritts Surveys")

The first large-scale aerial surveys for cetaceans in the Gulf of Mexico were conducted in 1979, 1980, and 1981 by the U.S. Fish and Wildlife Service (USFWS). These provided the first data on the distribution of offshore cetaceans in the Gulf. The usefulness of the information provided by the Fritts surveys, however, was limited by several factors. First, the surveys were conducted in only five limited Gulf of Mexico blocks, placed mostly over the continental shelf. Only off the Texas coast did the survey blocks include a large amount of continental slope and oceanic waters. Second, the small sample sizes precluded the possibility of making density estimates for most species.

The third and perhaps most important problem was with the species identifications. For example, the two species of spotted dolphins (*Stenella attenuata* and *S. frontalis*) were, at the time, not properly distinguishable (their taxonomy was not resolved until 1987). Thus, all spotted dolphins were lumped together, even though we now know that each of these two species is easily recognized. The same is true of the spinner (*S. longirostris*) and Clymene (*S. clymene*) dolphins, which were lumped together in the surveys. The striped dolphin (*S. coeruleoalba*) was reported as common off the west coast of Florida in shallow waters of 25 to 100 m or 328 ft. Because striped dolphins generally are considered deepwater animals, it seems possible that Fritts and co-workers misidentified some of these sightings. Finally, the five reported sightings of common dolphins (*Delphinus delphis*) from the pilot study in 1979 are questionable. There is currently no solid evidence that common

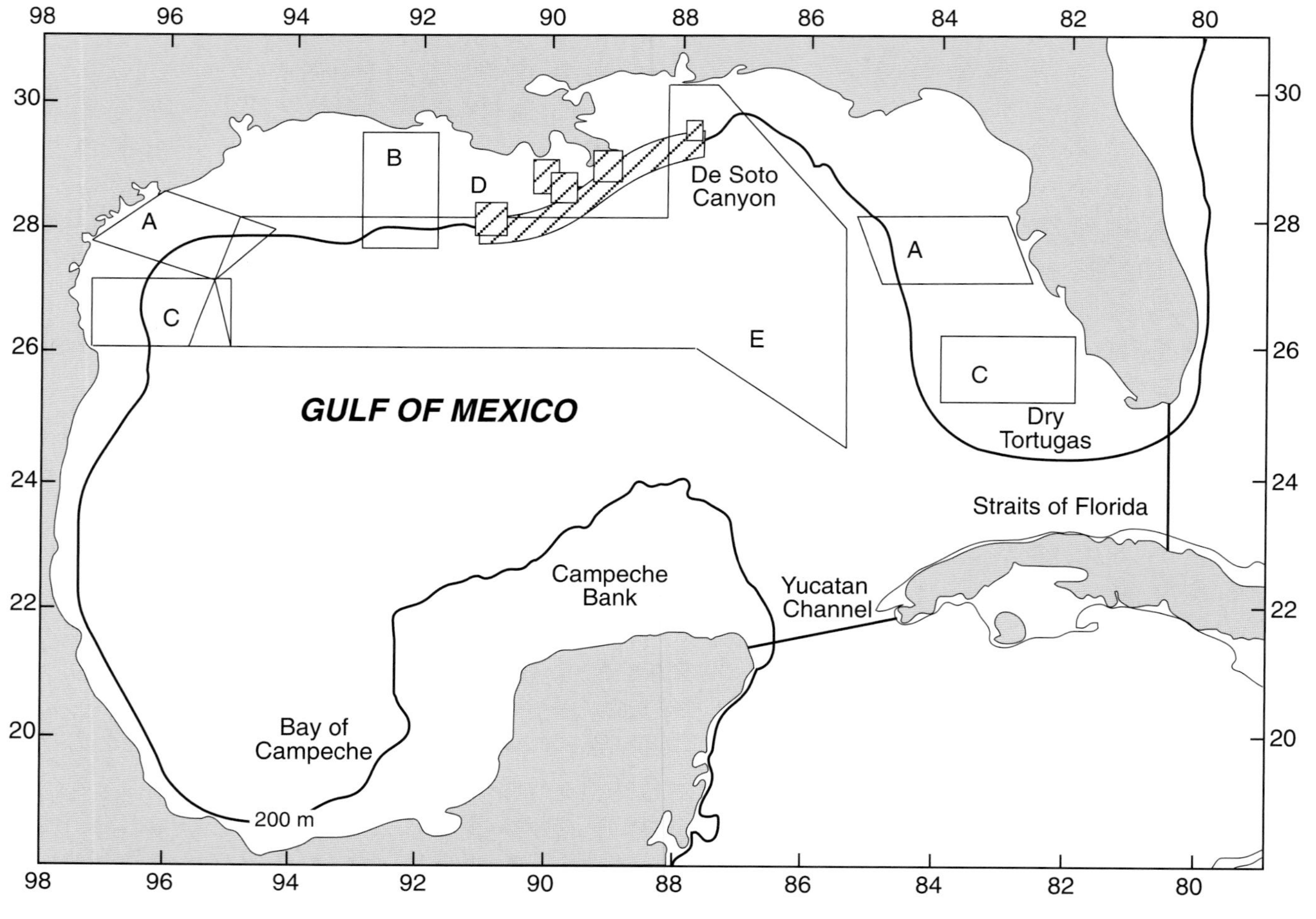
GULF OF MEXICO
De Soto Canyon
Dry Tortugas
Straits of Florida
Yucatan Channel
Campeche Bank
Bay of Campeche
200 m
A
B
C
D
E
North Latitude, in degrees
West Longitude, in degrees
98
96
94
92
90
88
86
84
82
80
30
28
26
24
22
20

*Figure 3. (left) Map of the study area showing locations of survey blocks from various studies. "A," "B," and "C" represent the Fritts surveys, "D" the CSAS surveys of 1989–90, and "E" the Oregon II surveys of 1990–94. The CSAS surveys are illustrated by angled bars within the survey areas. Numerous other boat and aerial surveys have been made in the shallow coastal waters. Courtesy Cartographic Service Unit, Department of Geography, Texas A&M University*



dolphins occur in the Gulf of Mexico; all previous records are either erroneous or questionable. It seems likely that early workers consistently misidentified some other species, perhaps *S. clymene* or *S. attenuata*, as common dolphins.

## NMFS Continental Shelf Aerial Surveys

The National Marine Fisheries Service (NMFS) conducted a set of aerial surveys from 1983 through 1986 to study the distribution of cetaceans in continental shelf waters of the Gulf. Bottlenose dolphins made up 97.6 percent of all cetaceans sighted; very few other species were observed.

## NMFS Red Drum Aerial Surveys

In 1986 and 1987, the NMFS conducted aerial surveys to estimate abundance of red drum (*Sciaenops ocellatus*) in the Gulf of Mexico. All efforts were in water less than 200 m, 656 ft., deep. Many bottlenose dolphin sightings were recorded, and used to compute abundance estimates for coastal waters; however, there were few sightings of other cetacean species.

## NMFS Continental Slope Aerial Surveys ("The CSAS Surveys")

Between 1989 and 1990, Mullin and others conducted aerial surveys for cetaceans in survey blocks off Louisiana and Mississippi,

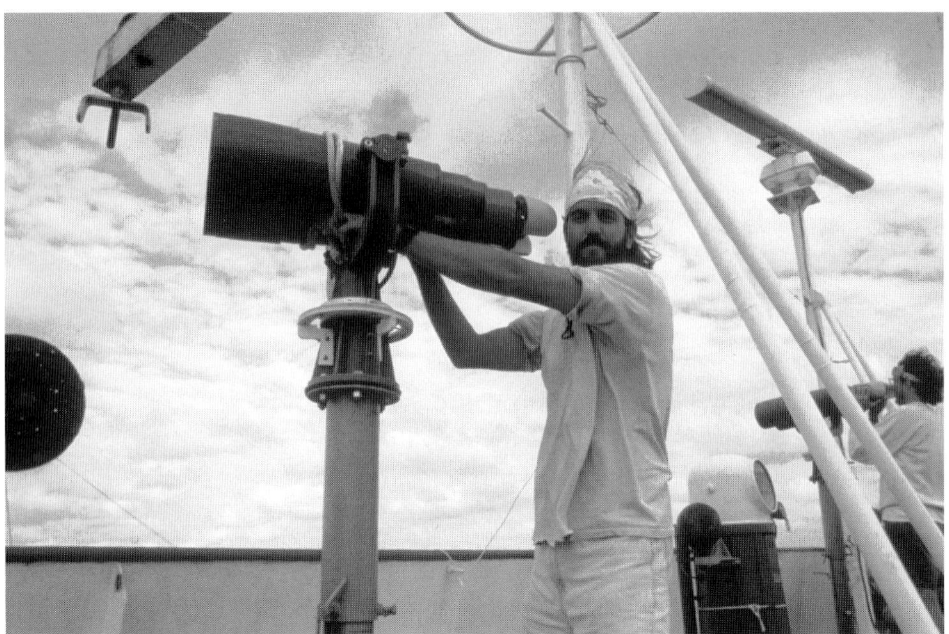

*33. David Weller of Texas A&M University is using Big Eye binoculars to scan for marine mammals on the high seas. These binoculars, generally used in tandem from the flying bridge of a ship, are the main tool for a type of survey termed "line transect" in which standardized methodology of data collection allows for estimation of population numbers from small subsamplings of the entire population.*

reported on in 1991. These provided excellent information on distribution and abundance of oceanic cetaceans in a restricted part of the Gulf of Mexico; however, like previous efforts, these were limited by their small-scale coverage.

## NMFS Vessel-Based Surveys ("The Oregon II Surveys")

Vessel-based offshore marine mammal surveys did not begin until recently. From spring of 1990 through 1994, the Southeast Fisheries Science Center of the NMFS conducted ship surveys for cetaceans in waters of the northern Gulf of Mexico. The transect lines included waters of the continental slope and all oceanic waters deeper than 100 m, or 328 ft., in the U.S. Gulf of Mexico. These surveys combined state-of-the-art survey techniques with highly qualified observers working in an extensive study area to provide the best information then available on cetaceans in the northern offshore Gulf of Mexico. These surveys have clarified several misconceptions about Gulf cetaceans. Their major limitation was that they provided information only about spring and early summer distribution patterns.

## Minerals Management Service/Biological Service Surveys (the "GulfCet Surveys")

In 1991, the Minerals Management Service (MMS) of the U.S. government contracted with Texas A&M University at Galveston (TAMUG) and the National Marine Fisheries Service (NMFS) to obtain detailed information on occurrence patterns and numbers of cetaceans and sea turtles in the area shown in Figure 4 (the northwestern and north–central part of the Gulf between the 100- (328 ft.) and 2000-m (6560 ft.) depth contours, or "on the continental slope"). These surveys consisted of at least 12 shipboard surveys and 8 aerial surveys covering all seasons. In conjunction with the spring–early summer "Oregon II Surveys" by NMFS, this intensive study presented us with much of the information that is summarized for "present status" of marine mammals in the deeper waters of the Gulf. Beginning in 1995, MMS teamed with the U.S. Biological Service to replicate these surveys, as well as to extend them farther eastward, to waters east of northern Florida

*Figure 4. Map of the northern Gulf of Mexico, showing the GulfCet I and II study areas, 1992–97. These represent the most comprehensive surveys in the Gulf accomplished at the printing of this book. Map by Don Frazier*

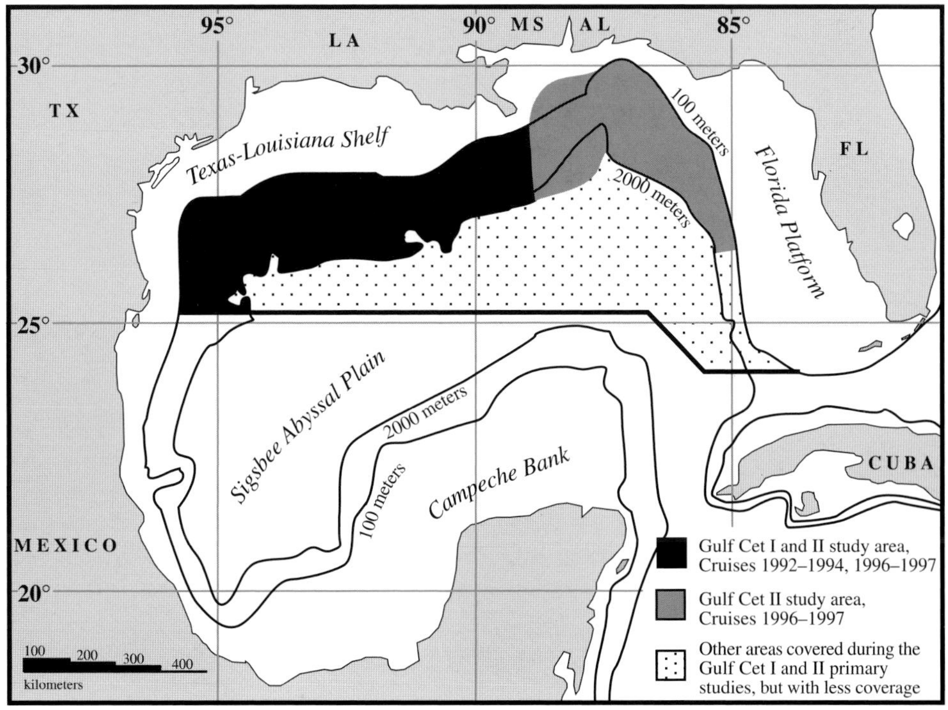

(200 m to 2000 m or 656 ft. to 6560 ft. deep). An especially prominent part of this latter work was the gathering of oceanographic and biological productivity data; the results helped to link occurrence patterns with some aspects of habitat, to be summarized for appropriate species in chapter 5.

# Chapter 5

# Accounts of Species

These descriptions of the biology and status of marine mammals known or suspected for the Gulf of Mexico rely to a large degree on information available from other areas of the world. For additional information on any of these species, we advise consulting the "general references" section of the bibliography.

## ORDER CETACEA, WHALES, DOLPHINS, AND PORPOISES

This order of wholly aquatic mammals occurs in all oceans and adjoining seas of the world, as well as in certain lakes and river systems. Of the thirty-four species of marine mammals recorded from or near the waters of the Gulf of Mexico, all but three are cetaceans, and all are oceanic forms.

Living cetaceans are traditionally divided into two suborders: the Odontoceti, individuals of which have teeth and an asymmetrical skull; and the Mysticeti, individuals of which have plates of baleen instead of teeth and a symmetrical skull.

### *Suborder Mysticeti, Baleen Whales*

Within the Gulf of Mexico, this suborder is represented by seven species arranged into two families, the Balaenidae and the Balaenopteridae.

#### FAMILY BALAENIDAE, RIGHT WHALES

This family, which includes three species, has the most extravagant baleen apparatus of all baleen whales. The head is correspondingly enormous, in some species up to almost 40 percent of the body length; there are no throat and chest grooves, as in the

Balaenopteridae, and no dorsal fin. The baleen plates are the longest in the Cetacea. A single species of this family is known from the Gulf of Mexico, and it is of extralimital distribution in the region.

*Eubalaena glacialis* Müller, 1776, Northern Right Whale



**Names.**—The generic name is from the Greek *Eu,* "good" or "right," and *balaena,* "a whale." The specific name is derived from the Latin *glaci,* "ice." Spanish: ballena franca.

**Description.**—Right whales are larger in girth relative to length than most other great whales, resulting in a rotund, "blimp-like" form, especially when compared to the streamlined rorquals. Average individuals are approximately 15 m (50 ft.) in length and weigh 54,500 kg (60 tons), although exceptionally large right whales may attain 18 m (60 ft.) and 90,900 kg (100 tons). Females are somewhat larger than males. The head is enormous—comprising about one-third of the total body length—and there are between 200 and 270 extremely long, dark gray baleen plates that hang from each side of the upper jaw (Table 1). The baleen plates may reach 3 m (9–10 ft.) in length.

The overall coloration of right whales is black or dark brown, mottled with white on the chin and belly. Right whales do not have the pleated throat characteristic of balaenopterid whales, and a dorsal fin is absent. The flippers are short and broad.

One of the most distinctive features of the right whale is the presence of numerous yellowish white skin growths called "callosities" along the lower jaw, the dorsal surface of the rostrum, and behind the blowholes. These crusty skin growths are composed of keratin, an ectodermal outgrowth similar in substance to hair and toenails of other mammals. They occur as soft bumps even in the fetal stage and continue to grow throughout life. The largest of these excrescences is positioned dorsally at the tip of the rostrum and is called the "bonnet." Some chin callosities contain hair follicles and are inhabited by cyamids, or "whale lice," and at times, barnacles. The size and patterns of callosities are distinctive and can be used to identify individual whales.

Right whales produce a wide, V-shaped blow when exhaling, and the tail flukes tend to show before deep dives. These whales have broad tail flukes that are deeply notched medially and are

**TABLE I.**
**Baleen plate dimensions for baleen whales occurring in the Gulf of Mexico.**

| *Species* | *Number of plates per side* | *Maximum length* | *Maximum width* | *Color* |
|---|---|---|---|---|
| Right whale | 200–270 | 280 | 30 | Dark gray to black |
| Blue whale | 270–395 | 100 | 55 | Black |
| Sei whale | 219–402 | 75 | 40 | Shiny black with fine white bristles |
| Fin whale | 260–480 | 90 | 50 | Gray; front 1/3 on right side all white |
| Bryde's whale | 250–370 | 45 | 25 | Dark gray with lighter gray bristles |
| Minke whale | 231–360 | 30 | 12 | Black with pale cream or white bristles |
| Humpback whale | 270–400 | 64 | 15 | Olive brown with gray bristles |

dark colored above and below. The trailing edge is not serrated, unlike the tail of the humpback whale.

**Behavior and Ecology.**—Right whales spend much of spring, summer, and autumn at high latitude feeding grounds and migrate to more southerly, warmer waters in winter for mating and calving. Due to the asynchronous seasons of the northern and southern hemispheres, the two species of *Eubalaena* (*E. glacialis* and *E. australis*) are not in the lower latitudes at the same time and interbreeding probably does not occur.

Right whales produce a variety of especially low-frequency, moanlike vocal sounds, as well as percussive sounds such as flipper slapping and tail slapping. A distinctive clacking sound has been described for right whales as they feed at the surface. Termed the "baleen rattle," this sound is produced by small wavelets rattling the baleen plates when they are held partially out of the water. Right whale sounds, both vocal and percussive, change with differing patterns of behavior and appear to have important communicative functions. Similar to other mysticetes, right whales do not echolocate with high frequencies like dolphins, but possibly may detect underwater features through reverberations of their low moans.

Right whales feed by skimming through concentrations of copepods and euphausiids (or "krill"), but sometimes may also take pelagic mollusks and juvenile red crabs. They have been described feeding in plankton slicks singly, in groups of 2 to 8 animals, and occasionally in loose aggregations of up to 30. The whales may skim through plankton at the surface and at depth, with plankton density rather than particular water depths determining where the whales feed. Right whales at times feed in coordinated fashion (echelon feeding) as they move forward in a V-shaped pattern, as the V of a goose formation, with all mouths open. In this manner, all but the lead whale take advantage of a wall of whale to one side, toward which prey cannot escape. Other baleen whales, including humpback whales (*Megaptera novaeangliae*), sei whales (*B. borealis*), and fin whales (*B. physalus*), feed with right whales at various times.

Reproduction in the southern right whale (*E. australis*) has been studied extensively in protected waters near the coast of Peninsula Valdés, Argentina. This prime winter and spring right whale mating and calving area has been repeatedly surveyed for whales by air and from the shore since 1970, and over 1000 individual whales have been identified by their distinctive head markings. Survey results indicate that mature females generally give birth at three- to seven-year intervals, and the age at first calving seems to be about seven years. Mothers with calves prefer water approximately 15-m (50 ft.) deep, and segregate spatially from the remaining whale population. The right whale population separates into three overlapping and nonexclusive groups: mothers and calves in one area, males and mature females without calves in a second, and a third group comprised of all sex and age groups, including mating pairs. Generally, mature females are seen in these waters only in years in which they give birth.

Reproduction in the northern right whale also has been studied, calving interval and segregation of mother–calf groups appears to be similar in *E. glacialis* as in the southern right whale. Waters off the southeastern coast of the United States are a primary calving ground for western North Atlantic right whales.

**Distribution and Status.**—Right whales are found in southern hemisphere oceans as well as in the North Atlantic and North Pacific, and are designated as two species in the Southern versus Northern Hemisphere. They have been decimated in all oceans, and only remnant populations exist. They are especially scarce in the North Pacific.

In the western North Atlantic, they are found from Iceland to Florida and as strays to the Gulf of Mexico, but their range may have been greater in pre-whaling days. In 1992, the minimum population number of western North Atlantic right whales was calculated as 295 animals. Worldwide, as few as 3000 to 4000 right whales may remain. This species is listed as endangered by the U.S. Fish and Wildlife Service (USFWS) and the International Union for the Conservation of Nature (IUCN).

**Status in the Gulf.**—Only two records are known from the Gulf of Mexico. On March 10, 1963, two right whales were observed off New Pass, Sarasota, Florida. These whales were swimming in water 9.4- to 10.3-m (30–40 ft.) deep and were approached by three observers as closely as about 4 m (12 ft.). One of us, David Schmidly, documented a stranded right whale on the Texas coast near Freeport, Brazoria County, Texas. Only a portion of the animal washed ashore, but the total length of the whale was probably 6 m (20 ft.) or less, indicating that the whale was a calf or young-of-the-year.

Although there are many other records from the east coast of Florida, this region represents the southern limit of the species' range in the western North Atlantic. The central Gulf of Mexico may have been a whaling ground for right whales in the 1880s, but nothing is known of this reputed whaling effort. Northern right whales are only of accidental occurrence in the Gulf of Mexico, and this is certainly one of the rarest of cetaceans in these waters.

**Remarks.**—Right whales were so named by early whalers because they were the "right" whale to kill in the sense of being slow swimmers that could be easily caught, usually floated when dead, and produced large quantities of oil and baleen. Consequently, right whales were decimated early by the world's whaling industries and have yet to recover.

## FAMILY BALAENOPTERIDAE, RORQUALS

This family is comprised of two genera and six species that occur in all oceans and adjoining seas of the world, with all six species having been recorded from the Gulf of Mexico. The term rorqual is derived from the Norwegian words *ror,* for "tube" or "groove," and *qual* for "whale." These whales vary enormously in size, with head and body length of sexually mature adults 6.7–31.0 m (22–102 ft.) and weights up to 160,000 kg (176 tons). In each species, the females are larger than the males of the same age. All of these whales are characterized by longitudinal folds or pleats on the throat, chest, and in some species, the belly. These furrows allow the throat to expand enormously and thus greatly increase the amount of material that the whale can take in when feeding. The Balaenopteridae are distinguished from the Balaenidae by the presence of throat and chest furrows, a more elongate and streamlined body form with relatively smaller head and more tapering pectoral fin, a softer and less massive tongue, and shorter and less flexible baleen. They also have a small dorsal fin.

*Balaenoptera musculus* (Linnaeus, 1758), Blue Whale

**Names.**—Derived from the Greek *balaena,* "a whale," and *pter,* "a wing," the generic name refers to the dorsal fin present in this genus of great whales. The specific name is from the Latin *muscul,* "muscle," for the size and power of the animal. Spanish: ballena azul.

**Description.**—Typified by enormous size, the blue whale is the largest of whales and is believed to be the largest creature ever to inhabit the earth. Blue whales commonly reach lengths of 24 to 26 m (80–85 ft.) and weigh approximately 100,000 kg (100 tons). Exceptional individuals may attain 30 m (100 ft.) in length and weigh 150,000–200,000 (150–200 tons). Like all baleen whales, females grow slightly larger than males of comparable age.

The blue whale has 55–88 deep ventral grooves, or "throat pleats," that extend from the lower jaw laterally along the belly to the navel. Blue whales have a large and rather broad rostrum (which is not as pointed as in other rorquals), and from the paired blowholes a single ridge extends along the dorsal surface of the head nearly to the end of the snout. The dorsal fin is small—only about 33 cm (13 in.) in height—and is located far back on the body. Blue whales are a light, bluish gray overall, mottled with gray or grayish white splotches. The flippers are light gray to white below and the baleen plates are black (Table 1).

When exhaling, blue whales show a tall, slender blow that is up to 9-m (30 ft.) high. This is followed by a view of the back, then often the small dorsal fin, and finally by broad, slightly raised tail flukes as the animal completes its dive. Commonly, a series of breaths are taken before a longer, 10–12 minute dive is accomplished.

The blue whale may be confused with the fin and sei whale at a distance; however, the latter two species differ in being gray above and white below with no mottling. Also, the fin whale is asymmetrically colored about the jaw, with part of the right side being white and the left side black. Both fins and seis have much larger dorsal fins per body size than blue whales.



**Behavior and Ecology.**—Spring and summer finds these whales migrating northward to subarctic feeding grounds. They move back to temperate waters in fall and winter where mating and parturition take place. The same trend is exhibited by blue whales in the southern hemisphere, although their feeding grounds are in the antarctic. As the seasons are reversed between the two hemispheres, northern and southern stocks do not normally interbreed in temperate and equatorial waters.

Formerly, it was thought that several baleen whales, including the blue whale, fed only on high latitude feeding grounds in the summer months and fasted for the seven- to eight-month period of migration and breeding. Although feeding seems to occur mainly in the higher latitudes, recent evidence has shown that these whales also commonly feed during migration. For example, in the North Pacific they feed off California every summer and autumn.

Small, shrimplike crustaceans known as "krill" predominate in their diet. In the southern oceans they exist almost exclusively on *Euphausia superba,* although smaller amounts of copepods and amphipods also are consumed. In the North Atlantic, they feed on several euphausiids, including *Meganyctiphanes norvegica, Thysanoessa inermis, T. raschii,* and *T. longicaudata.* Tremendous amounts of these small organisms are required to sustain a single whale. An adult must consume approximately 3000–5000 kg (3.0–5.0 metric tons) of krill per day to meet its energy requirements.

Blue whales feed by lunging into heavy concentrations of krill, or "plankton slicks," and gulping tremendous volumes of water. Up to 64,000 kg (64 metric tons) of water may be engulfed in just one mouthful. This action causes a spectacular distention of the throat area, made possible by the unique morphology of the ventral throat pleats, and changes the streamlined, cigar-shaped profile of the animal to a bloated, almost tadpolelike appearance. The water is then forced through the baleen plates, which act like a sieve to retain the krill, and the baleen is presumably scraped by the tongue to facilitate swallowing. Blue whales often lunge at the surface in tandem, apparently using each others' bodies as walls.

Low-frequency tonal sounds of more than 20-second duration have been recorded from blue whales, and probably are used in communication. These low-frequency "moans" last about 30 to 40 seconds and range in frequency from 12.5 to 200 Hz. These sounds are repeated again and again, interrupted only when the animal surfaces to breathe. They carry up to several hundred miles through ocean waters, depending on local conditions, and in all likelihood, along with those of fin whales, are the most powerful biologically produced sounds. In addition to communication, the moans may aid in navigation by serving as a form of long-distance echolocation for oceanographic features.

Females give birth to a single calf in temperate or equatorial waters during the winter months. Gestation is approximately 11 months and females bear young at 2- to 3-year intervals. Although the birth event has never been observed, whaling records indicate that a newborn blue whale is about 7 m (23 ft.) in length and weighs 2000–3000 kg (2–3 metric tons). Weaning occurs about seven months after parturition when the juvenile whale is approximately 12.8 m (42 ft.) in length. Sexual maturity is reached at about five years of age. At this time males are approximately 22.6-m (74 ft.) long and females are slightly larger at 24.0 m (79 ft.). Their life span is not known, but it is certainly greater than 50 years for both females and males.

**Distribution and Status.**—Blue whales are distributed worldwide. One subspecies, the pygmy blue whale (*B. m. brevicauda*), occurs in the southern hemisphere and at least in the northern part of the Indian Ocean.

These whales were heavily exploited during the whaling boom years and as a result are now uncommon. Pre-whaling stocks were estimated at 166,000 to 226,400 animals worldwide, but by the early 1970s, only 11,000 to 12,000 remained. The National Marine

Fisheries Service estimates the world's population of blue whales at approximately 11,700, with about 10,000 of these in southern oceans, 1600 in the North Pacific, and only low hundreds in North Atlantic waters. A major aggregation is in the Gulf of St. Lawrence, where 308 individuals have been identified as a minimum population. Despite their protected status, blue whale stocks have yet to show an increase in numbers over the past decade, and they are listed as endangered by the USFWS and the IUCN.



**Status in the Gulf.**—Only two reliable reports are available from the Gulf of Mexico, and both are of stranded animals. On August 17, 1940, a 21-m (70 ft.) blue whale stranded on the Texas coast between Freeport and San Luis Pass. This animal was alive when found and identified from photographs. A stranding of a purported fin whale near Sabine Pass, Louisiana, in early December 1924, was subsequently identified as a blue whale. Blue whales are probably only of accidental occurrence in the Gulf of Mexico and remain one of the rarest of cetaceans known from these waters.

*Balaenoptera physalus* (Linnaeus, 1758), Fin Whale

**Names.**—The specific name is from the Greek *physalis,* "a kind of toad which puffs itself up," probably in reference to the spectacular distension of the throat cavity during feeding lunges. Another common name is finback. Spanish: rorcual común.

**Description.**—The fin whale is second in size only to the blue whale. Northern hemisphere adults commonly reach maximum lengths of 24 m (79 ft.) and maximum weights of 50,000 kg (50 metric tons), with females tending to be slightly larger than males. Southern hemisphere animals are even longer, ranging up to 27 m (85 ft.) in length. Fin whales are somewhat slimmer than blue whales of comparable length, and therefore weigh proportionately less. They also have a head ridge extending medially from the blowhole to the tip of the snout, and a narrow rostrum that appears more pointed than that of the blue whale which it resembles. The dorsal fin, located farther from the tail flukes than in the blue whale, is taller and may reach a height of 61 cm (24 in.). The fin whale has 50–100 long throat pleats that reach back to the navel.

Fin whales have the distinction of being the only marine mammal with asymmetrical coloration. On the right side, much of the lower jaw and baleen plates are white, whereas on the left side, the baleen and jaw are nearly black. Overall body color is dark gray above and white below, without the mottling of blue whales. Ventral portions of the tail flukes and flippers are also variably white. A light V-shaped pattern, or "chevron," occurs on the back just behind the head.



In surfacing, fin whales exhibit a 5- to 8-m (15–25 ft.) blow and rarely lift their tail flukes above the water surface. They also show a peculiar humping to the back when diving, especially during slow movement. This occurs as the caudal peduncle (the portion of the animal just anterior to the tail) rises above the water surface but the tail itself does not.

**Behavior and Ecology.**—Fin whales are highly migratory, moving to high latitude feeding areas during spring and summer, and returning to lower, temperate water latitudes for mating and calving during autumn and winter. Like other migratory baleen whales, northern and southern hemisphere populations probably do not interbreed in warm temperate and tropical waters due to asynchronous seasons.

Fin whales feed mainly on krill, but also eat schooling fishes, including herring, cod, mackerel, pollock, sardine, and capelin. It is possible that for some populations, fish are eaten more often in winter. Fin whales have been observed feeding on schooling fishes with humpback whales, but the two species contrast in feeding style. Fin whales approach the fish horizontally and at slower speeds than the humpbacks, which rush the fish from below. Other whales feeding in the same vicinity may include right whales and sei whales. Direct competition among sympatric species is probably avoided by a generally different prey selection, although some overlapping may occur. Larger rorquals and right whales seem to feed predominantly on crustaceans and copepods, whereas the smaller rorquals include a larger percentage of fishes in their diet.

Mating and calving occur from November to March in northern latitude temperate waters. The gestation period is approximately 11 months and newborn fin whales are about 6.4-m (21 ft.) long and weigh nearly 1800 kg (18 metric tons). The suckling period lasts about seven months, and after weaning the young whales are approximately 11.5 m (38 ft.) in length. Sexual maturity is reached at 6 to 10 years of age, when males are 19 m (62 ft.) and females are 20 m (66 ft.) in length. Females are thought to give birth only at three-year intervals.

**Distribution and Status.**—Fin whales occur in all oceans of the world. There are two, possibly three, separate populations in the western North Atlantic: a northern, cold-adapted and a more southerly stock.

The worldwide, pre-whaling population has been estimated at 422,200 to 474,700 animals. The worldwide stock is about 106,000 to 122,000 animals, with approximately 3600 to 6300 of these in the western North Atlantic. Although fin whales are one of the most abundant of baleen whales, they are not overly common in the North Atlantic and perhaps never were. The USFWS lists them as endangered.



**Status in the Gulf.**—Fin whales have stranded in the Gulf on five occasions; there have been another seven sightings but only three can be confirmed. Sighting and stranding records have been made throughout the year, and these are the second most frequently reported baleen whales from the Gulf, after Bryde's whales. Adequate data do not exist for reliable population estimates, however, and it is likely that fin whales are simply accidentals from outside the Gulf. Bryde's whales are more common in these waters.

### *Balaenoptera borealis* Lesson, 1828, Sei Whale

**Names.**—The specific name is derived from the Latin *borealis,* "north" or "northern." Spanish: rorcual del norte.

**Description.**—The sei whale (pronounced "say") is a medium-sized baleen whale that typically reaches lengths of 15 to 16 m (49–52 ft.). Females are usually a little larger than males of comparable age and may attain maximum lengths of 20 m (66 ft.). Average weight is 13,000–15,500 kg (14–17 tons) with a probable maximum of about 29,000 kg (32 tons).

Sei whales are dark, bluish gray in coloration with an irregular white patch in the area of the ventral pleats. No white markings occur on the flippers, but the flanks and belly are mottled with random patterns of oval-shaped, light grayish to whitish marks that may represent scars left by lampreys. The dorsal fin is tall, up to 61 cm (24 in.) in height, strongly curved or hooked, and situated farther forward on the body than in blue or fin whales. The throat pleats are shorter than those of most other rorquals and extend to just beyond the flippers. Sei whales rise horizontally to breathe, with the head, dorsal fin, and much of the back coming into view at once. They normally do not dive deeply, and submerge by quietly slipping below the surface rather than arching the back and flinging the tail above water like deeper diving whales. These whales are similar in color and size to the Bryde's whale, and the two are extremely difficult to distinguish at a distance. Positive identification can only be made by close inspection of the head; sei whales generally have the single head ridge typical of other balaenopterids, whereas Bryde's whales have three head ridges (Plate 11).



**Behavior and Ecology.**—Like other baleen whales, sei whales are highly migratory. Calving and mating occur during the winter in tropical and subtropical waters. In spring and summer, they move to high latitude feeding grounds, but rarely penetrate far into polar regions.

Unlike most other balaenopterids, which mainly feed by engulfing huge amounts of water and food at one time, sei whales often skim their food from the water surface. A sei whale has been observed feeding with right and fin whales off Cape Cod, Massachusetts. In contrast to the other whales, the sei whale fed within 1 to 2 m (3–7 ft.) of the surface and seemed to ignore schooling fishes in favor of plankton slicks. Sei whales, however, are known to consume fishes, including anchovies, sauries, mackerel, capelin, cod, and sardines. Squid, copepods, euphausiids, and amphipods also are eaten.

Sei whales are usually encountered in small groups of between two and five animals, but may be found in larger congregations in summer feeding grounds. Adult females give birth to a single calf at 2- to 3-year intervals following a gestation period of 11 to 12 months. Newborns suckle for approximately six months and are sexually mature at eight years. Length at sexual maturity for females is 13.3 m (43.6 ft.) and 12.9 m (42.3 ft.) for males.

**Distribution and Status.**—Sei whales are widely distributed in all oceans of the world but are uncommon in most tropical regions. In winter months, they may be found in warmer waters. In summer, they disperse toward polar regions, but not as far into polar waters as several other rorquals. Difficulty in distinguishing sei and Bryde's whales at sea has hampered efforts to determine sei whale distribution in tropical waters, where the Bryde's whale is prevalent.

In the western North Atlantic, sei whales are found mainly in offshore waters from the Gulf of Mexico and Caribbean Sea northward to Nova Scotia and Newfoundland. Two, possibly three, discrete populations may occur in these waters, perhaps including a small Caribbean and Gulf of Mexico stock.

Pre-whaling populations numbered 172,700–197,700 animals worldwide. The U.S. National Marine Fisheries Service estimates the current world's population at 48,000 to 63,000. The western North Atlantic stocks total about 2600. Sei whales are listed as endangered by the USFWS.



**Status in the Gulf.**—Only five reliable records and one questionable one are available from the Gulf of Mexico. Three of the five reliable records are from strandings in eastern Louisiana, and one is from the Florida panhandle. Population estimates and natural history information on these whales in the Gulf are not available.

*Balaenoptera edeni* Anderson, 1879, Bryde's Whale

**Names.**—The specific name, *edeni,* is in honor of Sir Ashley Eden, who was Chief Commissioner of British Burma, where this whale was first described. Spanish: rorcual tropical.

**Description.**—Bryde's whales are small rorquals that average approximately 13 m (43 ft.) in length and 12,000 kg (12 metric tons) in weight. Maximum length is near 15.5 m (51 ft.). Like all other balaenopterids, females tend to be slightly larger than males.

The overall coloration is variable, but usually the dorsal surface is a dark, bluish gray or bluish black with the ventral region white or yellowish. The dorsal fin reaches a maximum height of 46 cm (18 in.) and is located about one-third of the body length forward of the tail flukes. The dorsal fin rises abruptly from the back, allowing for distinction from the fin whale, in which the forward edge of the dorsal fin juts more gradually from the back. There are 40–70 ventral pleats extending to the navel.

The most important distinguishing character of Bryde's whales is the presence of three head ridges extending along the dorsal surface of the head from the blowholes to the tip of the rostrum (Plate 11). All other whales of the genus *Balaenoptera* have only one such head ridge.

Bryde's whales are very similar in appearance to the sei whale, which is only slightly larger. A clear view of the head, allowing for determination of the presence of one or three ridges, is required to distinguish between these two species and positively identify either whale at sea.

**Behavior and Ecology.**—There is an unfortunate paucity of natural history and population data for the Bryde's whale, making this one of the least known of the great whales. This is primarily because even in the 1970s, whalers did not distinguish between Bryde's and sei whales.



These whales are usually seen singly or in small groups, although larger concentrations may be found in areas of food abundance. They feed largely on schooling fishes, including pelagic fishes such as pilchard, mackerel, herring, mullet, and anchovies, but also eat cephalopods and pelagic crustaceans.

Bryde's whales have a lengthy breeding period and may even breed year-round in some regions. As a result, birthing also occurs throughout the year; however, there is evidence for peaks in calving for some populations during winter and early spring. The period of gestation is approximately 12 months, and they give birth every other year. The calf is weaned after six months at a length of 7.1 m (23.3 ft.). Age at sexual maturity is 8–11 years, when females are approximately 12 m (39.4 ft.) in length and males are 11.9-m (39 ft.) long.

**Distribution and Status.**—Bryde's whales appear to be limited to tropical and warm temperate waters of the Atlantic, Indian, and Pacific Oceans and are not found in colder, high latitude waters. Some tropical populations may be nonmigratory, whereas those in temperate waters make only limited movements to warmer waters in winter and to more temperate regions in summer.

Few reliable population data are available for these whales. The worldwide population had been estimated at 30,200 to 55,500 animals, but no firm census figures are available for North Atlantic waters. The USFWS does not list this whale as either endangered or threatened.

**Status in the Gulf.**—Twelve verified stranding reports and 12 confirmed live sightings are available from the Gulf of Mexico, making this the most commonly observed baleen whale of the Gulf. All of the strandings are from the northeastern section of the Gulf between the Mississippi Delta and southern Florida. The live sightings are from every season but fall in the DeSoto Canyon region and off western Florida. Most whales were sighted in relatively shallow water near the 100-m (328-ft.) isobath.

Although stranding records and live sightings still are limited observations, the stranding reports are from all seasons and seem to indicate that Bryde's whales are present in the Gulf of Mexico throughout the year. Estimated abundance in the northern Gulf of Mexico is presently about fifty whales, but this is certainly the most numerous baleen whale in these waters, and this estimate may rise with further studies.



*Balaenoptera acutorostrata* Lacépède, 1804, Minke Whale

**Names.**—The specific name is from the Latin *acutus,* "sharp," and *rostrum,* "a beak" or "snout," referring to the sharply pointed snout of this species. Spanish: rorcual enano.

**Description.**—Averaging only 7–8 m (23–26 ft.) in length, this is the smallest of the rorquals and the second smallest of the baleen whales; the smallest is the poorly known pygmy right whale *Caperea marginata* of the southern hemisphere, which reaches a maximum length of only 6.4 m (21 ft.). The maximum recorded length for a mature female is 10.7 m (35 ft.) for southern hemisphere minkes and 9.2 m (30 ft.) for specimens from the North Atlantic. Maximum weight is 9100 kg (9.1 metric tons).

As its scientific name implies, the minke whale is distinguished by a very narrow and pointed rostrum. The triangular-shaped head has a single median head ridge extending anteriorly from the blowholes to the tip of the snout. The dorsal fin, located in the latter third of the back, is tall and falcate (sickle-shaped). The ventral throat pleats, numbering between 30 and 70, are short and end just behind the slender and sharply pointed flippers.

Overall coloration is dark gray to black dorsally and white ventrally. The undersides of the flippers are white, and in the northern hemisphere, the dorsal surface of the flippers has a prominent white band. Although a valuable diagnostic character, this white flipper band is variable in extent and may be absent in some Atlantic specimens. Regions of light gray coloration may extend up the flanks from the ventral region, and a light gray chevron is sometimes present dorsally, just behind the head. The baleen plates are yellowish white or cream colored.

Minke whales are difficult to locate and identify at a distance, as the blow is small and inconspicuous. On occasion, however, they closely approach boats where their relatively small size, and the combination of a sharply pointed rostrum, tall and sharply curved dorsal fin, and white flipper bands readily distinguish them.

**Behavior and Ecology.**—Minke whales frequently are observed singly or in small groups, although large numbers may be seen in areas of food abundance. They often segregate by sex and age in the nonbreeding period (summer), with males migrating farther north than females and immatures.

They feed on euphausiids and copepods, squid, and a variety of fishes, including sand lance, sand eel, salmon, capelin, mackerel, cod, coal fish, whiting, sprat, wolffish, dogfish, pollock, haddock, and herring. Capelin are the dominant food item of North Atlantic minkes and spawning concentrations of these fish apparently influence minke distribution in these waters.

Minke whales produce diverse sounds, including a variety of short and irregular "grunts, pings, zips, ratchets, and clicks."

In the North Atlantic, mating occurs from October to March. Following a gestation period of approximately 10 months, females give birth to a single calf in early winter, possibly as often as annually. Newborns are 2.4–2.7 m (8–9 ft.) in length and the period of lactation is four to six months. Approximate age and length at sexual maturity is 6 years and 6.75 m (22 ft.) for males and 7.1 years and 7.15–7.50 m (23.5–24.6 ft.) for females.

**Distribution and Status.**—Minke whales are cosmopolitan in distribution, occurring in all oceans of the world. Three geographically isolated groups occur in North Pacific, North Atlantic, and southern hemisphere waters.

Some populations display pronounced migratory tendencies, moving to cold temperate and polar waters in spring and back to warmer waters in autumn. In the northwest Atlantic, these whales are seasonally abundant and range from the Davis Strait of Baffin Bay during the summer to the Florida Keys and West Indies in winter. In contrast, populations off Puget Sound, Washington, appear not to migrate, and individually identified whales frequent this area as year-round residents.

Their distribution in the North Atlantic is heavily influenced by spawning concentrations of capelin upon which they feed. Also, seasonal segregation by sex and age is pronounced in the North Atlantic. Mature males migrate farther north in spring and summer than females and immature whales.

Current worldwide population estimates are between 315,800 and 331,800 animals. The average number estimated from the Gulf of Maine and lower Bay of Fundy is 2650 whales. Minkes are not listed by the USFWS as either endangered or threatened.



**Status in the Gulf.**—Most reports of minke whales from the Gulf of Mexico have come from the Florida Keys, although strandings in western and northern Florida, Louisiana, and Texas have been reported. Strandings are from winter and spring months. On December 30, 1953, a minke whale stranded alive approximately 0.3 km (0.5 mi.) east of Little Duck Key in the Florida Keys. According to eyewitnesses, it was pursuing a school of fish when it lodged itself on a sandbar and subsequently was shot by an onlooker.

On March 29, 1988, an immature female stranded on Matagorda peninsula of the Texas coast. Apparently, this whale came close to shore four to five days before it stranded alive. A necropsy revealed no obvious injuries or illness, but a small amount of plastic was found in the stomach. The physical features of this specimen were particularly interesting in several respects. First, the white bands usually present on the dorsal surface of the flippers were not evident in this whale. Also, the baleen plates were partially black; the usual color is grayish to cream white. This was the first such stranding of a minke whale on the Texas coast and is the westernmost record for the species in the Gulf of Mexico.

All ten confirmed records are of strandings, and no minke whales have been seen alive in the Gulf. Because of the winter–spring nature of the strandings, these animals may represent a northward migration of whales from the open ocean or the Caribbean Sea.

*Megaptera novaeangliae* (Borowski, 1781), Humpback Whale

**Names.**—A reference to the large flippers of this whale, the generic name is from the Latin *mega,* "great," and *pter,* "a wing." The specific name, *novaeangliae,* is Latin for "New England." Spanish: rorcual jorobado.

**Description.**—Humpback whales typically reach lengths of 15 to 16 m (49–52 ft.) and weigh between 31,000 and 41,000 kg (31–41 metric tons). The maximum recorded length is 18 m (59 ft.). Relative to other balaenopterid whales, they are greater in girth and present a bulkier, less streamlined appearance.

Coloration is black overall with irregular white markings on the throat, sides, and abdomen. In some individuals the belly may be entirely white, or there may be white patterns dorsally. The flippers, which extend to 4.6-m (15 ft.) long, are white ventrally, but on the dorsal surface range from black to patterns of black-and-white, or are entirely white. The tail flukes are broad, serrated on the trailing edge, and generally black above but display highly variable black-and-white color patterns on the ventral surface, ranging from all black to nearly all white. The distinctive tail flukes have been used successfully to identify individual whales.

A unique characteristic is the presence of numerous knobby structures, or "dermal tubercles," on the median head ridge, dorsal surface of the snout, chin, and mandible. The number and location of these bumps vary among individuals, and each tubercle contains a sensory hair.

The dorsal fin, which may be up to 31 cm (12 in.) in height, is falcate to rounded in profile and is located about one-third of the body length from the tail flukes. The fin is situated on a hump that is accentuated when the whale arches its back in diving.

These whales have between 14 and 35 ventral throat pleats that reach to the navel or just beyond. These are fewer and spaced wider apart than typical for balaenopterids. This, together with the stockier body shape, slight median head ridge, enormous flippers, and the numerous dermal tubercles of the head, lead most cetologists to separate the humpback from the "true" rorquals.

Humpbacks typically submerge for 6 to 7 minutes at a time with occasional dives of 15 to 30 minutes. The blow may be up to 3 m (10 ft.) high, and the plume is dispersed and bushy in appearance. When diving, humpbacks arch the back steeply, which accounts for their common name. The flukes rarely show before shallow dives, but when a deep dive is started, they are usually lifted high above the water surface. Since the flukes are variably tattered and have white markings on the underside, they have been used extensively for individual identification.

**Ecology and Behavior.**—Humpbacks often congregate in scattered groups of 20 to 30 individuals, but have been observed in larger congregations of 100 to 200 animals. These are among the most acrobatic and visible of whales and often breach completely out of water in spectacular displays of strength. They commonly slap their tail flukes or flippers on the water's surface and occasionally lift their huge heads above water to peer about, a behavior known as "spyhopping." Tail slapping, breaching, and other such behaviors may serve in communication between the dispersed whales, possibly as warnings or a means of indicating location.



Humpbacks also produce a number of unusual sounds described variously as moans, groans, cries, squeals, chirps, and clicks. Sounds are arranged into complex and predictable patterns analogous to human and bird songs. These songs may be repeated for long periods of time, and most often have been recorded in low latitude breeding grounds. Songs are broadcast by sexually mature, lone males and are most likely used as either intra-sexual or intersexual (or both) signals associated with mating.

These whales eat krill, mackerel, sand lance, capelin, herring, pollock, smelt, cod, sardines, salmon, and anchovies. They are lunge feeders, using several different techniques to concentrate their food before lunging. An especially interesting technique is known as "bubble-netting." In bubble-netting, one or more humpbacks exhale while circling below a food source. The resulting bubble column is thought to effectively form a net to concentrate food items before the whales lunge upward with open mouths.

Females give birth to a single calf in tropical or subtropical waters during winter. The gestation period is approximately 11 months. Newborn humpbacks are 4.6 m (15 ft.) in length and weigh about 1400 kg (1.4 metric tons). The period of lactation lasts approximately five months. Sexual maturity is reached in two to five years, at which time the young whales measure about 12 m (39 ft.) in length. Females breed every two to three years, but may do so with only one-year spacing as well.

**Distribution and Status.**—Humpback whales occur in all oceans of the world. In the western North Atlantic, they are distributed from north of Iceland and west of Greenland southward to Venezuela and around the tropical islands of the West Indies. Population estimates indicate that the worldwide, pre-whaling population was approximately 100,000 animals. The present stock numbers 9500–10,000, with about 5800 of these occurring in the western North

Atlantic. Humpback whales are listed as endangered by the USFWS and the IUCN.

Humpbacks are highly migratory. In the western North Atlantic, they occupy high latitude feeding grounds from Cape Cod to Iceland during spring, summer, and fall. In late autumn and winter, the whales move into Caribbean waters for mating and calving.



**Status in the Gulf.**—In the Gulf of Mexico, humpback whales occasionally were hunted near the Florida Keys, but are uncommon in the Gulf proper. Sightings have been made off the west coast of Florida and near Alabama in the eastern Gulf, and off the jetties in Galveston, Texas, in the western Gulf. On May 5, 1997, six humpbacks were sighted in a single group, about 250 km (155 mi.) east of the Mississippi Delta at a depth of 1000 m (3,280 ft.).

Their social sounds have been heard with underwater microphones, or hydrophones, on at least three occasions in the northwestern part of the Gulf. In addition to sightings, two strandings of humpbacks have been documented. In December 1932, a humpback stranded 12.4 km (20 mi.) from Havana, Cuba, and in March 1983, one stranded alive near Seahorse Key, Levy County, Florida. A resident population of these whales apparently does not occur in the Gulf of Mexico, and it is likely that they are accidentals from the Caribbean. It has been suggested that as these whales recover from past intensive hunting, increasing numbers of juveniles will spend their winters in the Gulf.

### *Suborder Odontoceti, Toothed Whales and Dolphins*

The vast majority of marine mammals in the Gulf of Mexico region are representatives of this suborder. There are 9 families and 68 species worldwide, of which 4 families and 24 species have been recorded from (or probably occur in) the Gulf of Mexico. The Gulf representatives include a wide variety of toothed whales and dolphins, ranging in size from the large sperm whales (body length: 15 m) to the small oceanic dolphins (body length: 150–250 cm).

#### FAMILY PHYSETERIDAE, SPERM WHALE

This family includes a single species, the sperm whale, which occurs in all oceans and adjoining seas of the world, except in polar ice fields. This is the largest toothed mammal in the world and the most sexually dimorphic of all cetaceans. The most striking morphological feature is the huge spermaceti organ in the head, filled with up to 1900 liters of waxy oil. The sperm whale is by far the most common large whale in the Gulf of Mexico.

*Physeter macrocephalus* Linnaeus, 1758, Sperm Whale



**Names.**—The generic name is from the Greek *physeter,* "blower," referring to the blowhole. The Greek *makros,* "large" or "long," and *kephale,* "head," refer to the enormous size of this whale's head. Another scientific name, *P. catodon,* is still widely used in the scientific literature. Spanish: cachalote.

**Description.**—The sperm whale is unique in appearance and the most familiar of the great whales. This is Herman Melville's "Moby Dick," and a mainstay of the United States' former whaling industries. This species has been popularized in literature and films and the one that the public identifies as "the whale."

The sperm whale exhibits a remarkable degree of sexual dimorphism in body size, with males noticeably larger than females. This situation is the reverse of that in baleen whales. Sperm whales are also the largest of the odontocetes; males average about 15 m (50 ft.) in length and weigh approximately 36,000 kg (36 metric tons), and females average 11 m (36 ft.) and about 20,000 kg (20 metric tons). Maximum size is believed to be about 20 m (65 ft.) for males and 17 m (56 ft.) for females.

These whales are characterized by an enormous, blunt head that is squarish in profile and comprises 25–35 percent of the total body length. The single, S-shaped blowhole is located at the front of the snout and is displaced to the left, rather than centered on top of the head. Therefore, when exhaling, they show a blow that is 3- to 5-m (10–16 ft.) high and directed to one side when seen from the front or rear of the animal. A side view shows the spout slanting forward at a 45° angle.

Sperm whales have 18–25 teeth in each side of the lower jaw (Table 2). The upper jaw also may have teeth, but these rarely erupt. Their teeth average 23 cm (9 in.) in length and are not covered with enamel; thus, they are easily worked with carving tools and were long a favorite of scrimshaw (pictures etched into teeth or ivory and then blackened with ink or soot) artisans. The mandible is greatly underslung and gives the appearance of being undersized for such a giant head.

**TABLE 2.**
**Dentition of odontocetes occurring in the Gulf of Mexico.**

| | NUMBER OF TEETH | | |
|---|---|---|---|
| *Species* | *Each side upper jaw* | *Each side lower jaw* | *Comments* |
| Sperm whale | 10–16 | 18–25 | Upper teeth rarely emerge, lower teeth large and fitting into sockets in upper jaw |
| Pygmy sperm whale | 0 | 12–16 | Very pointed |
| Dwarf sperm whale | 0–3 | 8–11 | Very pointed |
| Blainville's beaked whale | 0 | 1 | Displaced 50% from tip of jaw |
| Gervais' beaked whale | 0 | 1 | Slightly displaced (15%) from tip of jaw |
| Sowerby's beaked whale | 0 | 1 | Displaced 50% from tip of jaw, crown directed posteriorly |
| Cuvier's beaked whale | 0 | 1 | At tip of jaw |
| Killer whale | 10–12 | 10–12 | Large, pointed, curved |
| Pygmy killer whale | 8–13 | 10–13 | Lower teeth smaller |
| False killer whale | 7–12 | 7–12 | Prominent, pointed, curved |
| Melon-headed whale | 20–25 | 22–24 | |
| Short-finned pilot whale | 8–9 | 7–9 | Curved |
| Rough-toothed dolphin | 20–27 | 20–27 | Tooth crowns with fine vertical wrinkles |
| Short-beaked common dolphin | 40–54 | 40–54 | Upper jaw with deep palatal groove |
| Long-beaked common dolphin | 47–60 | 47–60 | |
| Bottlenose dolphin | 20–26 | 18–24 | |
| Risso's dolphin | 0 (1–2 rarely) | 0–7 | Near front of jaw |
| Atlantic spotted dolphin | 30–42 | 30–42 | |
| Striped dolphin | 35–55 | 35–55 | |
| Pantropical spotted dolphin | 34–48 | 34–48 | |
| Spinner dolphin | 45–65 | 45–65 | Very fine, sharply pointed |
| Clymene dolphin | 38–49 | 38–49 | |
| Fraser's dolphin | 38–44 | 38–44 | |

Overall coloration is dark, brownish gray to bluish black, except for areas of white around the lower jaw and belly. The whales are usually mottled with numerous scars and grayish in color; the skin is creased with wrinkles, producing a shriveled appearance. Cases of partially or entirely white sperm whales have been authenticated. The dorsal fin is replaced by a large hump and is followed by several smaller humps to the flukes, creating an irregular ridge along the dorsal surface of the tailstock. Sperm whales have stubby flippers and broad, powerfully wielded tail flukes.

These whales spend 10–20 minutes breathing at the surface before making deep dives that last 60 minutes or more. When diving, the tail flukes are nearly always thrown high, giving the observer a good view of triangular-shaped flukes that are naturally smooth along the rear (or trailing) edge but develop nicks and notches with time and are darkly pigmented both dorsally and ventrally.



**Behavior and Ecology.**—Sperm whales are known to make deep, prolonged feeding dives that may extend to the ocean floor, even in water as deep as 3200 m (10,500 ft.) or more. Most feeding dives, however, descend only to about 1000 m (3,280 ft.). The long breathing time at the surface suggests that sperm whales may enter into lactic acid debt, whereby they metabolize oxygen that must be replenished later.

At such tremendous depths, these remarkable animals hunt their primary prey, squid. Up to 910 kg (0.91 metric tons) of squid per day are required to sustain a single sperm whale. Besides squid, they consume a wide variety of other deepwater prey, including octopus, lobsters, crabs, jellyfish, sponges, and numerous varieties of fishes. The range in size of prey items, from giant squid measuring 3 m (9.84 ft.) along the mantle to lantern fish only a few centimeters (one inch or two) in length, may be the greatest of all living mammals. This circumstance, coupled with the realizations that no light penetrates to the depths at which these animals feed and that squid are highly elusive swimmers, has led to much speculation concerning the feeding method of sperm whales.

They may catch their food by: lying suspended and relatively motionless near the ocean floor and ambushing prey; attracting squid and other prey with bioluminescent mouths; ingesting food with a sucking motion of the tongue; or stunning prey with ultrasonic sounds. Sperm whales occasionally suffocate after becoming entangled in deep-sea cables that wrap around their lower jaw, and odd objects (e.g., stones, rubber boots, buckets, and boards) have been found in their stomachs. Both circumstances suggest that the animals may at times cruise the ocean floor with open mouths.

Sperm whales produce a variety of click sounds that can evidently be focused and projected with great intensity, indicating the important role of sound in feeding. Interestingly, examination of stomach contents reveals little evidence that the whale's teeth and lower jaw are used to procure, grasp, or chew prey. Blind sperm whales, and others with dramatically deformed lower jaws, occasionally have been captured by whalers. These animals showed no apparent ill effects from their handicaps, lending indirect support to the idea that they somehow immobilize their prey before consumption. Perhaps sperm whales investigate their surroundings with ultrasound, and when prey items are discovered, intensify and beam the click sounds to stun the prey. Although an intriguing hypothesis, it has not been proven.

The role of sound is crucial in the life of these whales. Sound probably is used not only in feeding but also for orientation and communication. Trains of click sounds are heard in repeated, stereotyped "codas," or short series of 3 to 4 or more clicks. Codas are usually about 0.5 to 1.5 seconds in duration, and repeated 2 to 6 or more times at variable intervals of a few seconds or minutes. Codas are unique to individual sperm whales and are probably used in communication.

Each click of a sperm whale is composed of a substructure of clicks only microseconds apart from each other (indistinguishable to our ears except as the major click). Apparently, the first sub-click is produced near the front of the head. It escapes the head and also bounces (or echoes) off the skull behind where it is produced. This "bouncing" happens several times, and the substructure of clicks is related to the size of the head. Since males grow much larger than females, and old males are larger than young ones, it is very likely that click structure advertises size (and age) to other males and potentially to females. If this is true, the clicks may send an aural equivalent of antlers in male deer (i.e., "I am larger [and tougher] than you are") and may serve as a display signal.

Breeding behavior is very similar to that of elephants. A single, dominant male briefly joins a group of females and offspring and "defends" the group against competing males. Smaller, and apparently not yet ready to mate males, form "bachelor" groups. Battles between rival males for access to a group of females can occur. Twenty to thirty females and young may live together in a matriarchal group with long-term bonds of affiliation; mothers and their female offspring may stay together for life.

The period of gestation is 15 to 16 months and lactation may last 7 to 13 years in some cases. Newborn sperm whales are approximately 4 m (13 ft.) in length and weigh nearly 910 kg (0.91 metric tons). Although twin calves are known, a single calf per female is the rule. The calving interval may take as long as five to seven years. Females sexually mature at about age 8, and males at ages 12–15. Males, however, do not mature socially until after age 25.



The sperm whale has been the mainstay of the pelagic whaling industry. Prior to the advent of modern technology (harpoon cannons, diesel-powered catcher boats, and massive factory ships), the hunting of sperm whales was a dangerous occupation. Sperm whales effectively fought back on occasion, and one even sank an American whaler, the *Essex,* in 1820. Reportedly, the ship rammed the whale, which in turn repeatedly rammed the ship with its large head. Only three of the crew lived to tell the story. Sperm whales not hunted are known to be docile and harmless to humans.

Sperm whales were hunted the world over for an array of valuable products. Other than whale oil for lamps and lubricants, these whales provided spermaceti (oil from the forehead) for high quality, smokeless candles and ambergris (a waxy byproduct of digestion), which was used as a binding substance in the manufacture of fragrances. Spermaceti is a fluid encased within the huge head and is responsible for the whale's common name; early whalers also saw a resemblance of congealed spermaceti to sperm. Spermaceti is believed to aid in the echolocation process by providing a medium similar to the density of seawater around the sound-producing organ. It has been hypothesized incorrectly that sperm whales can control their buoyancy by cooling and heating the spermaceti in their heads.

Ambergris is a sebaceous secretion that forms around indigestible matter, especially cephalopod beaks, in the whale's stomach and intestine. This substance is common in sperm whales, but has been documented in fewer than 5 percent of other whales.

**Distribution and Status.**—Sperm whales are close to worldwide in distribution and occur in all oceans, including arctic and antarctic waters. They are found primarily in temperate and tropical waters of the Atlantic, Pacific, and Indian Oceans. In the western North Atlantic, they often are found along the continental shelf but rarely over the shelf itself, as they usually feed in deeper waters.

Prior to concentrated whaling activity, sperm whales numbered more than 2 million individuals worldwide, although there are no good estimates of the numbers that inhabited the North Atlantic. Approximately 22,000 sperm whales are believed to roam the North Atlantic with as many as 900,000 animals worldwide.



Sperm whales are highly migratory, especially the males. Adult males move into high latitude temperate and polar waters during summer and lead a solitary lifestyle, whereas female groups remain in tropical or subtropical waters. In winter, the bulls return to lower latitudes for mating. The USFWS lists the sperm whale as endangered.

**Status in the Gulf.**—Sperm whales are by far the most numerous large whale in the Gulf of Mexico, and at one time were hunted in these waters. During aerial and shipboard surveys during 1979–81, 47 adult and 12 young sperm whales were observed. Most of these sightings were at the continental shelf edge (200 m or 656 ft.) or over the slope, and 71 percent of the sightings occurred off the Texas coast.

Strandings have occurred in Texas, Louisiana, Florida, and north of Veracruz, Mexico, and these whales have stranded or have been sighted in the Gulf of Mexico in every month of the year. These factors suggest that a stock of sperm whales unique to the Gulf of Mexico may exist, but this remains to be substantiated. Information from the GulfCet project indicates that there are between 300 and 400 sperm whales in the northwestern Gulf of Mexico. They were sighted (and their clicks heard via hydrophones) in all parts of the northwestern Gulf of Mexico between the 100- and 2000-m depth contours; however, most animals were concentrated around the 1000-m depth contour south of the Mississippi River Delta, and at similar depths roughly 300 km (186 mi.) east of the Texas–Mexico border. Although the animals occur there year-round, details of potential seasonal movements are unknown, and it is not clear if other concentrations of sperm whales occur in the Gulf.

Recent evidence suggests that sperm whales south of the Mississippi River Delta adjust their movements to stay in or near a variable area of upwelling cold water, or cold-core ring. If true, this presumably is related to higher productivity in this dynamic, continually shifting, oceanographic regime.

The stranding of an infant sperm whale near Sabine Pass, Texas, on September 2, 1989, made national headlines. Estimated to have been no more than two weeks old, the 545.5 kg (1200 lb.), 4-m (12 ft.) long whale stranded alive and was transported by truck to Sea-Arama in Galveston. Despite the efforts of staff and volunteers, the whale, which had been nicknamed "Odie" and followed closely by the popular press, died one week later from an apparent lung infection.



FAMILY KOGIIDAE, PYGMY AND DWARF SPERM WHALES

This family includes two species of the genus *Kogia,* both of which occur commonly in the Gulf of Mexico. Among the cetaceans, the facial part of the skull in these whales is among the shortest. These whales resemble the giant sperm whale (*Physeter*) by having a spermaceti organ in the head and functional teeth confined to the lower jaw. For these reasons, previous authors typically combined the pygmy and dwarf sperm whales into a single family with the giant sperm whale. Both species of *Kogia,* however, have blowholes situated on top of the head instead of at the end of the snout, and a distinct, curved dorsal fin; *Physeter* has no true dorsal fin.

*Kogia breviceps* (Blainville, 1838), Pygmy Sperm Whale

**Names.**—The origin of *Kogia* is unknown, but the word may be a Latinized form of the term "codger," for a "miserly old fellow," and its use has been attributed to a Turk named Cogia Effendi who was a Mediterranean whale observer. The specific name is from the Latin *brevis,* "short," and *cephitis,* "head," referring to the relatively short head. Spanish: cachalote pigmeo.

**Description.**—In profile, whales of the genus *Kogia* show a blunt, squarish head with a narrow and underslung lower jaw, although this feature is not as pronounced as the massive, rectangular profile of the giant sperm whale (*Physeter macrocephalus*). The head in *Kogia* constitutes only about 15 percent of the total body length versus approximately 30 percent for the sperm whale, which gives pygmy and dwarf sperm whales a somewhat sleeker head shape that has been described as sharklike in appearance. Also, these whales are markedly smaller than the giant sperm whale, with adult pygmy sperm whales reaching lengths of only 2.7 to 3.7 m (9–12 ft.) and weights of 317 to 410 kg (700–900 lb.).

In contrast to the larger sperm whale, the blowhole of *Kogia* is located on top of the head above the eye rather than at the end of the snout, and these smaller whales possess a well-developed and distinct dorsal fin that has no resemblance to the lumpy median ridge running along the tailstock of the sperm whale. The dorsal fin in *K. breviceps* is small and positioned posterior to the center of the back.

Coloration in the pygmy sperm whale is dark gray dorsally, fading to light gray on the sides and dull white to pinkish ventrally. A crescent-shaped mark of lighter pigmentation called the "bracket" or "false gill" is located in the area between the eye and flipper. There are 12–16 (sometimes 10) teeth in each side of the lower jaw, but the upper jaw usually has no teeth (Table 2).

Characters useful in distinguishing *K. breviceps* from the strikingly similar *K. simus* are detailed in the account for the latter.

**Behavior and Ecology.**—One result of the nomenclatural confusion surrounding the genus *Kogia* is a critical lack of reliable information on the life history of these whales. Observations of pygmy and dwarf sperm whales were often recorded with no distinction made between the two species, and very little is known of the food habits, reproductive cycle, and population dynamics unique to each species.

Rarely seen at sea by casual observers, these whales most commonly occur in small groups of three to six individuals and are rather slow moving and deliberate in their actions. They tend to rise slowly to the surface to breathe, produce an inconspicuous blow, and do not roll and splash about at the surface like many other smaller cetaceans. Also, when startled they often defecate a cloud of reddish brown or chocolate colored feces before diving.

Low-frequency, low intensity pulsed sounds have been recorded from a stranded animal in California, suggesting that these whales are capable of echolocation. Pygmy sperm whales feed mainly on squid, but also eat crab, shrimp, and smaller fishes.

Their reproductive habits remain poorly known. The gestation period is 9–11 months and peak calving is from autumn to spring. A newborn is about 1.2-m (4 ft.) long and 54 kg (120 lb.) at birth. The stranding of lactating and pregnant females in the company of newborn calves suggests that calving may at times occur in successive years.

**Distribution and Status.**—Pygmy sperm whales are distributed worldwide in warm ocean waters and have generally been regarded as relatively rare, although they commonly are stranded. Their apparent rarity may be due in part to the difficulty of observing the animals at sea.

In the western North Atlantic, these whales are known as far north as Sable Island and Halifax, Nova Scotia, as far south as Cuba (and probably farther south), and as far west as Texas in the Gulf of Mexico. Pygmy sperm whales most often are seen over and near the continental slope. Frequent strandings may be associated with movements into nearshore waters during calving; however, seasonal migrations are unknown. There are no population estimates for these whales.

**Status in the Gulf.**—Pygmy sperm whales frequently strand on the coastline of the Gulf of Mexico. These whales are most likely a common element in the nearshore Gulf fauna, but their habits prevent them from being readily observed. Many strandings appear to be related to the birth process; females with newborn calves often strand, as do recently postpartum females.

During 1984 to 1990, the Southeastern United States Stranding Network, a cooperative program of the Division of Mammals at the Smithsonian Institution and the Cousteau Society, documented the strandings of 22 pygmy sperm whales along the Gulf of Mexico coastline. Fifteen of these records were from Florida, six from Texas, and one from Mississippi. Interestingly, although this whale is a year-round resident of these waters, most strandings occurred in the autumn and winter months. Seasonal movements within the Gulf of Mexico remain undocumented.

During the 1992–97 GulfCet aerial surveys in the northwestern part of the Gulf, pygmy and dwarf sperm whales were lumped because of difficulties in distinguishing the species from the air. They were sighted in all waters between the 100- and 2000-m (328–6,560 ft.) depth contours. Group sizes averaged about 1.5 to 2.0 animals and ranged from 1.0 to 6.0. Densities were highest in spring and summer and much lower in the fall and winter. Estimate of abundance from shipboard and aerial sightings in all seasons was about 1,000 individuals in the northern Gulf.

**Remarks.**—Due to the general similarities in skull structure that dwarf and pygmy sperm whales share with their giant relative, the sperm whale, early cetacean taxonomists included these whales in the genus *Physeter.* This nomenclatural controversy finally began to resolve itself by the early twentieth century when the two species of pygmy sperm whales, *Kogia breviceps* and *K. simus,* were described. These two species are so similar in appearance, however, that this arrangement generally was not accepted until the 1960s. They were placed in the family Physeteridae with the sperm whales until recently when the family Kogiidae was recognized as a separate taxonomic entity.



*Kogia simus* (Owen, 1866), Dwarf Sperm Whale

**Names.**—The specific name is from the Latin *simus,* "flat-nosed." Spanish: cachalote enano.

**Description.**—The dwarf sperm whale is a slightly smaller version of the pygmy sperm whale, *Kogia breviceps,* reaching lengths of 2.1 to 2.7 m (7–9 ft.) and weighing 136–212 kg (300–500 lb.). Like *K. breviceps,* the dwarf sperm whale has a blunt, stubby head with a small and underslung mandible, a light colored "false gill" located just behind the eye, and dark, bluish gray or "steel gray" dorsal coloration that blends to lighter shades on the sides and to white ventrally. The dorsal fin is well developed, falcate, and somewhat taller than in *K. breviceps.* The height of the dorsal fin is greater than 5 percent of the total body length in *K. simus,* whereas it is less than 5 percent in *K. breviceps.* The dorsal fin is positioned farther forward than in the pygmy sperm whale, and thus appears more centrally located on the back.

In stranded specimens, examination of the lower jaw will readily distinguish the two species. Dwarf sperm whales have 8 to 11 (rarely 12–13) teeth in each side of the lower jaw, whereas pygmy sperm whales have 12 to 16 teeth. Also, the mandibular symphysis (where the two bones of the lower jaw fuse between the front teeth) of dwarf sperm whales measures 37–46 mm (9.4–11.7 in.) in length and has no ventral keel, or ridge of bone along the lower surface. In the pygmy sperm whale, the mandibular symphysis is 86- to 120-mm (21.8–30.5 in.) long and keeled ventrally. Dwarf sperm whales also have several small creases in the skin of the throat area that are lacking in pygmy sperm whales; these may be visible on stranded specimens.

The unique body form of the genus *Kogia* easily identifies these small whales at sea if a good view of the head is obtained; however, distinguishing between the two species can be done only by experienced observers.

**Behavior and Ecology.**—Despite fairly frequent strandings, very little definitive natural history information on the dwarf sperm whale is available. These whales usually are seen in small groups of fewer than five animals, and groups may be segregated by sex and age. They are thought to make deep and prolonged dives in pursuit of squid, fishes, and crustaceans.



The reproductive habits are almost completely unknown. Calves are believed to measure approximately 1.1 m (3.5 ft.) in length at birth and weigh about 45.3 kg (100 lb.). The young reach sexual maturity when they are about 2.1 m (7 ft.) in length.

**Distribution and Status.**—Dwarf sperm whales are distributed in warm temperate and tropical oceans of the world. In the western North Atlantic, they are known from Virginia to the Lesser Antilles and the Gulf of Mexico. Their range appears to overlap that of *K. breviceps.*

Dwarf sperm whales strand fairly frequently, although not as often as pygmy sperm whales, and are known to inhabit deep water in offshore areas. Strandings in shallow waters may be associated with the calving period. Seasonal movements, although they may occur, have not been documented; while no population estimates exist, this whale does not appear to be rare.

**Status in the Gulf.**—Like pygmy sperm whales, stranding records of dwarf sperm whales from the Gulf of Mexico are more abundant than would be expected for a truly rare animal; these whales, however, are less frequently stranded than their pygmy relatives. From 1984 to 1990, the Southeastern United States Stranding Network documented 10 stranded dwarf sperm whales in the Gulf of Mexico. Seven of these records were from Florida, two from Texas, and one from Louisiana. Earlier stranding records also showed that this whale had stranded in Mississippi. Both pygmy and dwarf sperm whales seem to strand more often in the eastern Gulf of Mexico; the dwarf sperm whale occurs about half as frequently as *K. breviceps,* with strandings throughout the year.

During the GulfCet I study, dwarf sperm whales were sighted 22 times by ship during spring and summer, and their abundance in the northwestern study area was estimated at 46 to 170 animals.

A synopsis of shipboard and aerial sightings from the 1992–97 GulfCet surveys for both pygmy and dwarf sperm whales is given in the account of the former.



FAMILY ZIPHIIDAE, BEAKED WHALES

This family is comprised of 5 genera and 19 species which occur in all the oceans and adjoining seas of the world. The vernacular name is derived from the long, narrow beak that forms a continuous smooth profile with the head in all forms known from the Gulf of Mexico. In most ziphiids, the teeth show strong sexual dimorphism, with males having one pair of usually large teeth in the lower jaw; in females, the teeth are absent or vestigial. The cetacean fauna of the Gulf includes two genera and four species of beaked whales that are among the least known of cetaceans in these waters. They usually remain well out to sea, avoid ships, and dive to great depths to secure cephalopods and fishes. The taxonomy of beaked whales is presently being revised; and with analysis of newly discovered skull fragments and newly acquired genetic samples, it is likely that several new species will be described in the next several years.

*Ziphius cavirostris* Cuvier, 1823, Cuvier's Beaked Whale

**Names.**—The generic name is from the Greek *Xiphias,* " a sword." The specific name is derived from the Latin *cavus,* "hollow," and *rostrum,* "a beak." Other common names include goosebeaked whale, Ziphius, and grampus. Spanish: zifio de Cuvier.

**Description.**—Female Cuvier's beaked whales grow slightly larger than males, which is typical for all beaked whales. Average length is 5.8 m (19 ft.) for females and 5.5 m (18 ft.) for males. Maximum length of females is about 7.0 m (23 ft.), and for males about 7.5 m (25 ft.). Maximum recorded weight is about 3000 kg (3.3 tons).

These whales have a relatively small head with a short and stubby, almost indistinct beak. The forehead slopes gradually to the beak and the lower jaw extends slightly beyond the upper. In profile, there is a slight concavity dorsally just behind the blowhole. The flippers are small and rounded, whereas the tail flukes are broad and at times have a small notch medially. The dorsal fin is falcate, up to 38 cm (15 in.) in height, and is located well behind the midpoint of the back. Ventrally, one pair of short grooves is apparent in the throat region. Males typically have a single pair of large, conical teeth that protrude from the tip of the mandible (Table 2). The teeth are up to 8-cm (3 in.) long, oval in cross section, and slant forward. As is typical of other ziphiids, females rarely have erupted teeth.

Coloration is highly variable and ranges from dark brown to slate gray dorsally with lighter shades below. The head is frequently pale except for two small, dark patches in which the eyes are located, giving the impression of a ring around the eye. Old males often are scarred extensively by parasites and encounters with other Cuvier's beaked whales, which inflict characteristic two-tracked scars with their two large teeth. Cuvier's beaked whales can be confused in particular with species of the genus *Mesoplodon.*

**Behavior and Ecology.**—Despite being among the more abundant and frequently stranded of the mysterious beaked whales, life history information on Cuvier's beaked whale is extremely sparse. These whales generally are observed in groups of 3 to 10 individuals, although large schools of 25 have been reported. Solitary bulls are observed occasionally. These whales are deep divers and may remain submerged for 30 minutes or longer while feeding on squid, fishes, crabs, and starfish.

There does not seem to be a distinct breeding season as calves are born year-round; the length of gestation is unknown. Calves are about 2.7-m (8.8 ft.) long at birth.

**Distribution and Status.**—Cuvier's beaked whale is widely (but sparsely) distributed throughout tropical and temperate oceans of the world. It apparently is absent from waters above 60°N and below 50°S. In the western North Atlantic, it has been recorded from Massachusetts and Rhode Island southward to Florida, the West Indies, and the Gulf of Mexico. This whale strands more frequently than other ziphiids, but no reliable estimates of abundance are available.

Cuvier's beaked whale is an offshore species. Seasonal movements remain to be substantiated, although general movements from tropical to temperate waters during the summer months may occur. This whale, however, appears to be a year-round resident of some parts of its range.

**Status in the Gulf.**—Eighteen strandings of Cuvier's beaked whales have been reported by the Southeastern United States Stranding

Network or are known from the historic record of Gulf cetacean strandings. Strandings have occurred in all seasons, with a slight peak in spring. Most strandings have been from the eastern portion of the Gulf, primarily from Florida.

During the 1992–97 GulfCet surveys, Cuvier's beaked whales were seen 10 times from the ship and 4 times from the air, in all seasons. These sightings were in the deepest part of the northwestern Gulf study area, around 2000 m (6,560 ft.) deep; the estimated abundance in that area was about 160 animals.



Unidentified ziphiids were sighted in all seasons 23 times from ships and 17 times from the air during the GulfCet surveys. Group sizes ranged from 1 to 7, with averages of 2.4 (from ship) and 3.2 (airplane). Unidentified ziphiids were sighted throughout the northwestern Gulf study area, but generally in waters 1000 to 2000 m (3,280–6,560 ft.) deep. Abundance for that area was estimated at about 200 animals.

*Mesoplodon densirostris* (de Blainville, 1817), Blainville's Beaked Whale

**Names.**—The generic name, *Mesoplodon,* is derived from three Greek words: *mesos,* "the middle," *hople,* "arm" or "armed," and *odon,* "a tooth." All refer to the tooth present in the middle of the lower jaw of most (but not all) members of this genus. The specific name is from the Latin *dens,* "thick," and *rostrum,* "a beak," an allusion to the dense bone comprising the beak of this whale. Another common name is the densebeaked whale. Spanish: zifio de Blainville.

**Description.**—Blainville's beaked whale reaches approximately 4.7 m (15.5 ft.) in length and weighs about 910 kg (1 ton). Like other members of this genus, this species is slender in body form, has short flippers that appear to be set low on the body, a falcate dorsal fin set far back on the body, and a slender and pointed rostrum, or beak. A single pair of throat grooves characteristic of the family Ziphiidae is present (but inconspicuous) in this species.

The beak and head shape are somewhat different from other ziphiids, at least in males. Mature males have a single large tooth located atop the midpoint of each mandibular side (Table 2). This tooth may be up to 20 cm (8 in.) in length and is embedded in a large hump of supporting bone that gives a high, arching contour to the jaw. Females have neither such prominent teeth nor a strongly crested profile to the lower jaw. The combination of a large tooth set at the midpoint of a highly arching lower jaw is diagnostic of this whale, and is sufficient to identify males of this species at sea if a good sighting of the head is obtained.

Coloration is blue–gray dorsally and white ventrally. Like most ziphiids, these whales are usually mottled and streaked with numerous grayish marks left by ectoparasites and squid during feeding; scars from wounds (probably inflicted by the slashing teeth of males during intraspecific fighting) are common. The tail flukes are not notched medially and may even bulge slightly in this area.

**Behavior and Ecology.**—Almost no natural history information is available for this secretive whale. It has been observed at sea singly and in small pods of three to seven animals. It is known to feed on squid, although fishes are taken at times.

Sounds recorded from a subadult male stranded near Crescent Beach, Florida, revealed several faint but distinct, rapidly pulsed "chirps" and "whistles."

The reproductive habits are almost completely unknown. Length at birth is thought to be 2.0–2.5 m (6.6–8.2 ft.), and age at sexual maturity is estimated to be nine years.

**Distribution and Status.**—This beaked whale is distributed throughout tropical and warm temperate waters of the world but is nowhere common. In the western North Atlantic, it occurs from Nova Scotia southward to Florida, the Bahamas, and the Gulf of Mexico, and is not as common as in some other regions. Although thought to be rare, no substantial population data exist for this unusual whale, and seasonal movements remain undocumented.

**Status in the Gulf.**—Only four verified stranding records are available from the Gulf of Mexico. These are from the Mississippi–Alabama border, the Atchafalaya Bay area of Louisiana, Matagorda Bay along the Texas coast, and the panhandle of Florida. Two Blainville's beaked whales have been sighted in the northern Gulf of Mexico. A synopsis of live unidentified ziphiids seen during the GulfCet study is presented in the account of Cuvier's beaked whale.

### *Mesoplodon bidens* (Sowerby, 1804), Sowerby's Beaked Whale

**Names.**—The specific name is from the Latin *bis,* "two," and *dens,* "a tooth." Another common name is North Sea beaked whale. Spanish: zifio de Sowerby.

**Description.**—Sowerby's beaked whale reaches a length of about 5.5 m (18 ft.) in males and at least 5.1 m (16.7 ft.) in females. Maximum weight is 1400 kg (1.5 tons). This whale is slender in body form with a relatively long, slender beak that slopes to a pronounced bulge just forward of the blowhole. The flippers are also relatively long and paddle-shaped.

Like other mesoplodont whales, this species is characterized by a unique arrangement of teeth in the lower jaw. Mature males have one pair of teeth set obliquely in the mandible that are approximately 10 cm (4 in.) in length and positioned at the midpoint of the mandible, about 30.5 cm (12 in.) from the tip (Table 2). The teeth have long roots and the tooth crowns point posteriorly. Females have smaller or unerupted teeth.

Coloration is charcoal gray dorsally, grading to light gray ventrally. These whales may become extensively mottled with light gray scars as they age.

**Behavior and Ecology.**—This is one of the most commonly stranded beaked whales, but little is known of its natural history and behavior. It feeds on squid and small fishes. Mating and calving probably occur from late winter to spring, and newborns are about 2.4 m (7.9 ft.) in length.

**Distribution and Status.**—Sowerby's beaked whale occupies cold temperate to subarctic waters of the North Atlantic and is the most northerly species of *Mesoplodon* in the Atlantic Ocean. In the western North Atlantic, this whale has stranded in Massachusetts, Newfoundland, and Labrador, but is more frequently reported from the eastern North Atlantic where it may be fairly common. In the western North Atlantic, there is one enigmatic stranding report from the subtropical waters of the Gulf of Mexico.

Seasonal movements are not documented, but these whales may make southerly migrations at the onset of winter to avoid advancing pack ice.

**Status in the Gulf.**—The only record from the Gulf of Mexico is of an animal that stranded alive in shallow water near St. Joseph Spit, Gulf County, Florida. A second small whale reportedly accompanied the stranded whale but apparently did not come ashore. The stranded whale died four days later while under veterinary care. This stranding represents the lowest reported latitude of occurrence for Sowerby's beaked whale and is approximately 2800 km (1739 mi.) from the previous southernmost North American record. It is doubtful that this beaked whale occurs regularly in the Gulf of Mexico; this unusual stranding probably represents no more than an extralimital record. Interestingly, a stranding also occurred in the Mediterranean Sea off Sicily, Italy.



*Mesoplodon europaeus* (Gervais, 1855), Gervais' Beaked Whale

**Names.**—The specific name refers to the English Channel, the type locality in which this whale was described. Other common names include the Gulfstream beaked whale and Antillian beaked whale. Spanish: zifio de Gervais.

**Description.**—This is one of the largest members of the family Ziphiidae. Mature females are 4.2–5.7 m (13.8–18.7 ft.) in length; males measure up to 4.6 m (15.1 ft.). Minimum weight is 1200 kg (1.3 tons).

Like other mesoplodonts, the dorsal fin is small, falcate, and located well behind the center of the back. The tail flukes are broad; the flippers are small and set low on the body. The head is relatively small and tapers rapidly to a narrow beak. In mature males, one pair of large teeth is set in the mandible about 7.6 to 10.2 cm (3–4 in.) from the tip of the snout (Table 2), about one-third the total length of the mandible.

Coloration is dark gray to black above and lighter ventrally, with occasional patches of white in the genital region. Frequently, there are numerous light gray splotches and scratches that contrast with the darker ground coloration.

**Behavior and Ecology.**—Almost nothing is known about the life history of this whale, although it is known to feed on squid. Strandings frequently are believed to be associated with calving, which may take place in shallow waters. Pregnant females or females with calves have stranded along the southeastern United States coast and the Gulf of Mexico in May, June, and August, suggesting a spring and summer calving period. Estimated calf length at birth is 2.1 m (6.9 ft.).

**Distribution and Status.**—Gervais' beaked whale is known from tropical, subtropical, and warm temperate waters of the Atlantic Ocean. Although it was described on the basis of a specimen found in the English Channel, no other specimen since has been recorded in European waters; the type specimen is regarded as a wide-ranging stray.

In the western North Atlantic, Gervais' beaked whale is known from Long Island, New York, southward to Florida and the Gulf of Mexico. It also has stranded in the Caribbean, western Africa, and on Ascension Island south of the equator.



**Status in the Gulf.**—Sixteen stranding records of this whale are available from the Gulf of Mexico, making it the most frequently stranded beaked whale in these waters. Four strandings are from middle to southern Florida, two from the northeastern Gulf, five from Texas, four from the northwestern shore of Cuba, and one from southern Mexico. This may be the most widely distributed ziphiid in the Gulf; however, this is not certain. For a discussion on sightings of live unidentified ziphiids, see the account of Cuvier's beaked whale.

### FAMILY DELPHINIDAE, OCEAN DOLPHINS

This is the most diverse family of cetaceans in the world, with a total of 17 genera and 33 species. It includes all of the classic dolphins, those smaller cetaceans with beaklike snouts and slender, streamlined bodies. Head and body length of delphinids in the Gulf of Mexico ranges from as little as 1.8 m (6 ft.) in species of *Stenella* to as much as 9.8 m (30 ft.) in *Orcinus;* weight in fully grown individuals ranges from about 50 to 10000 kg (about 100 lb. to 10 metric tons). The blowhole is located well back from the tip of the beak or the front of the head. The pectoral and dorsal fins are falcate, triangular, or broadly rounded, and the dorsal fin is located near the middle of the back. The Delphinidae include the most agile and some of the most speedy cetaceans. They commonly surface several times a minute and frequently leap clear of the water.

Abundance estimates for the GulfCet I study area (Figure 4) have been derived for many of the delphinids in the Gulf and are given (when available) in the "Status in the Gulf" section of species descriptions. The estimates refer to 95 percent confidence intervals derived from mathematical extrapolations of dolphins sighted by standardized surveys from ships and airplanes. Some updated estimates are presented for the GulfCet II study which also includes a northern Gulf area to the east of GulfCet I. Note that these estimates are only for a particular area in the northern Gulf and are certain to be refined in the future.

*Orcinus orca* (Linnaeus, 1758), Killer Whale

**Names.**—The scientific name is from the Latin *orca,* "a whale." Spanish: orca.



**Description.**—Killer whales are the largest member of the dolphin family Delphinidae. Maximum length for adult males is approximately 9.8 m (30 ft.); females reach up to 8.5 m (26 ft.). Females are less stocky than males and weigh up to 7500 kg (7.5 metric tons); large males may reach 10,000 kg (10 metric tons).

These whales are among the most easily recognized of cetaceans and often can be approached at close range. The body is stocky in form with a blunt snout and small, indistinct beak. The flippers are large and paddle-shaped and the dorsal fin is falcate to canted and exceptionally tall. In males, the dorsal fin may be up to 1.8 m (6 ft.) in height. Killer whales are black dorsally with a pale gray area called the saddle located just behind the dorsal fin. The size, shape, and color tone of the saddle varies not only between individual whales but also has characteristic forms for geographically separate stocks. Ventrally, killer whales are white from the chin to just behind the anus. The white ventral coloration extends up the sides posterior to the dorsal fin, and an oval patch of white is located just above and behind the eye. Although occasional melanistic or albinistic individuals have been encountered, this distinctive black-and-white color pattern, together with the large dorsal fin and white oval spot close to the eye, allows easy identification of this whale at sea.

Stranded whales can be identified by their large size and characteristic coloration (if still evident) as well as by the teeth. Killer whales have 10–12 slightly recurved teeth in each side of both jaws (Table 2) that are up to 13-cm (5 in.) long, oval in cross section, and interlock when the jaws are closed.

**Behavior and Ecology.**—Killer whales usually are observed in pods of 5 to 20 animals, although up to 250 have been seen together at one time. Large groups probably represent several smaller pods that have temporarily aggregated. Pods themselves appear to be very stable, with little or no immigration or emigration over many years. These whales are highly cooperative and the pod functions as a unit when hunting, making them extremely efficient predators. Pods usually contain adults of both sexes but females and young sometimes will segregate to form their own groups.

Prey items include squid, fishes, skates, rays, sharks, sea turtles, seabirds, seals, sea lions, walrus, dugongs, dolphins and porpoises, belugas, narwhals, and large whales such as fin, humpback, bowhead, right, minke, and gray whales. Pods have been observed attacking even sperm and blue whales. Killer whales often attack such large whales by hanging onto the fins and flippers, thereby immobilizing their prey; they also may attack the lower jaw and tear out the tongue. On the Atlantic coast of South America, as well as on islands in the Indian Ocean, killer whales have been seen lunging through the surf and coming onto the beach in pursuit of elephant seals and sea lions. After an attack, they then slide and wriggle back into water depths adequate for swimming. In captivity, killer whales eat about 45 kg (about 100 lb.) of food per day but free-range animals probably consume more.

Killer whales are proficient hunters and will take practically any palatable prey, but attacks on humans are extremely rare. Such an attack directed at one of us (BW) was effected by a killer whale that probably mistook its potential prey for a sea lion on the beach. Hunting techniques frequently are characterized by highly developed group coordination, and require good communication abilities. These whales are known to make a variety of sounds and are probably proficient at echolocation. Different pods have unique sound dialects, suggesting that sounds are passed on by oral tradition and individually modified from the basic genetic template for sound production.

Like most cetaceans, their reproductive habits are poorly known. The mating system tends toward polygyny, with one or several males having major access to reproductively active females; mates are probably outside of the males' immediate pods, but this remains to be confirmed. In Norwegian coastal waters, mating activity occurs throughout the year with a peak from October through December; the calving interval is about one calf every three years. Females reach sexual maturity at 4.6 to 4.9 m (15–16 ft.) in length and about 8 years of age; males are not sexually mature until 5.6 m (19 ft.) in length and 15 years old. Physical maturity in males is probably not reached until 20 to 25 years of age, and these whales have life spans of well over 50 years.

**Distribution and Status.**—Killer whales are distributed throughout all oceans of the world, including polar waters. In the western North Atlantic, they are known from the polar ice pack southward to Florida, the Lesser Antilles, and the Gulf of Mexico, although they are more common in the cooler waters from New Jersey northward. They seem to prefer shallow coastal waters, bays, and estuaries, but also occur in deepwater areas.



Seasonal movements are tied more closely to food availability than to water or climatic conditions. Killer whales arrive annually off the coast of New England with migrating schools of tuna. Along the eastern Canadian coast, the whales move inshore in spring and summer, and in autumn apparently move southward ahead of newly forming ice; however, this movement is not well documented. Killer whales are more commonly encountered in the Caribbean during summer than in other months.

**Status in the Gulf.**—Stranding records from the Gulf of Mexico are few and poorly documented. These include one possible stranding from the northern coastline of Cuba, one unverified report from south Texas, and three records from the Gulf coast of Florida.

Sightings in the Gulf of Mexico have occurred frequently in recent years. While only 9 reliable records were collated of live sightings before 1990, another 14 have been added since then, largely due to northwestern Gulf surveys associated with the GulfCet project. Each sighting included 1–12 whales observed during spring and summer in a relatively small area 200 to more than 2000 m (656–6,560 ft.) deep southwest of the Mississippi Delta. Earlier sightings also occurred in the summer, and this could relate to more fishing activity in deep water during the warmer months. There are many "consistent" summer reports (usually, however, without exact dates and positions) of killer whales from near the 200-m (656-ft.) depth contour off South Padre Island, Texas, and it is likely that the animals are frequent visitors to this area. Estimates from the GulfCet study indicate that about 70 killer whales occur between the 100- and 2000-m (328–6,560 ft.) depth contours. It is not known whether these animals stay within the confines of the Gulf or range more widely.

*Globicephala macrorhynchus* Gray, 1846, Short-finned Pilot Whale

**Names.**–The generic name is derived from the Latin *globus,* "a ball" or "globe," and the Greek *kephale,* "head." The specific name is from the Greek roots *makros,* "large," and *rhynchos,* "a snout" or "beak." Other common names include blackfish, pothead, pilot whale, and shortfin pilot whale. Spanish: calderón de aletas cortas.

**Description.**—The short-finned pilot whale is large and stocky, or "robust," in body shape. Males are much larger than females, reaching 6.1 m (20 ft.) in length; adult females measure up to 5.5 m (18 ft.). Maximum weight is 3600 kg (3.6 metric tons) for males and 2500 kg (2.5 metric tons) for females.

These whales have a bulbous, globular-shaped head, but contrary to their scientific name, do not have a beak; however, slightly protruding lips may give the impression of a beak. The bulbous head structure, or melon, is most highly developed in males and gives a blunt, squarish profile to the head. In old males, the melon may overhang the mouth.

In addition to the characteristic shape of the head, these whales also have a well-developed dorsal fin located forward of the center of the back. The fin is strongly curved and long at the base, but not very tall. The fin of older males is large, usually recurved strongly toward the back, and very thick. The flippers are long, slender, and sharply pointed at the tips. The tailstock is strongly compressed laterally and heavily keeled above and below.

Coloration is dark gray to black with a lighter gray patch on the chin and belly. This ventral patch is variable in shape but has been described as "anchor-shaped." There is a gray to white "inverted chevron" located dorsally just behind the head. The apex of this marking points toward the dorsal fin and is helpful in identifying animals from the air. A variably white saddle patch is present just behind and under the dorsal fin. Seven to nine short, robust teeth are present in each side of the upper and lower jaw toward the front third of the mouth (Table 2).

The short-finned pilot whale closely resembles the long-finned species (*G. melas*) which is frequently encountered along the northeastern U.S. coast. Although both species have flippers that are long and slender, the flippers are slightly shorter and straighter in the short-finned variety. Long-finned pilot whales, however, have not been recorded as yet in the Gulf of Mexico; in these waters, the combination of large size, dark coloration, markedly bulbous head, and distinctive dorsal fin distinguish the short-finned pilot whale from other whales and dolphins.

**Behavior and Ecology.**—These whales are highly gregarious social animals commonly encountered in groups of 10 to 60, although larger herds of several hundred are not infrequent. They often strand, and reports of mass strandings are numerous throughout warm waters of the world.



Short-finned pilot whales produce a variety of sounds, including both clicks and whistles used in communication and echolocation; a pod of pilot whales presents quite a cacophony of noise to underwater listening devices.

Their food habits are not well known; they are thought to feed mainly on squid, although fishes are also taken. Pilot whales also have been seen "harassing" sperm whales and dolphins on several occasions, and it is possible that they sometimes supplement their diet by eating marine mammals.

Breeding and calving are variable by stock, and may occur in winter, spring, or fall. Gestation lasts approximately one year. Calves are about 1.4-m (4.5 ft.) long at birth and weigh approximately 60 kg (130 lb.). Females are thought to give birth only once every three years. These whales live in stable, female-based matrilineal societies; postpartum lactation may indicate that older mothers are taking care of young that are not their own.

**Distribution and Status.**—Short-finned pilot whales are common residents of offshore tropical, subtropical, and warm temperate waters. They are known in the western North Atlantic from Virginia southward to northern South America, the Caribbean, and the Gulf of Mexico. Seasonal movements are not well documented, but seem to show a more southerly distribution during the spring and late winter in the western North Atlantic.

**Status in the Gulf.**—Short-finned pilot whales have been reported from the Gulf during all months of the year. Although apparent seasonal movements have been reported for the nearby Caribbean, no pattern of seasonal migration is evident for the Gulf region.

Fifteen mass strandings of more than five animals per stranding were recorded in the Gulf as of 1990. Of this total, 40 percent (6) were during July and August; however, mass strandings are known in all months except January and December. All but one of the mass strandings occurred in Florida. The exception was a mass stranding of 49 pilot whales on the coast of Louisiana in 1939 immediately following a severe hurricane.



A total of 64 reliable records of short-finned pilot whales are available from the historical record prior to the 1992–97 GulfCet program. Fifty were of strandings from the Florida Keys and the Florida west coast; the remainder were from Louisiana, Texas, and the Yucatan peninsula in southeastern Mexico. During the GulfCet surveys, these whales were sighted 15 times by ship and 12 times by air during all seasons; group sizes ranged from 2 to 85 animals. Sightings were mainly in the central and western (no eastern sightings) part of the GulfCet study area, generally in water 200 to 1000 m (656–3,280 ft.) deep, on the continental shelf slope. Abundance has been estimated at about 1700 animals in the entire northern Gulf.

Although short-finned pilot whales generally are considered common offshore residents of the Gulf of Mexico, stranding records have declined dramatically over the past decade. By comparing recent stranding records to historic records, one of us (DS) documented a 42 percent decline for the Gulf of Mexico and southeastern U.S. coast. Pilot whale reports comprised 7.1 percent of all cetacean strandings in the historic record (prior to 1978) from this region, whereas these whales comprise only about 4.1 percent of the modern record (1978 to present). These figures should not be cited as evidence of a steep decline in population numbers, but neither should they be ignored. Given the increased awareness of cetaceans in the past two decades, coupled with the appearance of stranding networks specifically organized to document reports of cetaceans, the modern record should show an increase in sighting and stranding reports if population levels remained stable. Thus, while certainly not conclusive, the available evidence points to the possibility of declining populations of these whales in the Gulf of Mexico and nearby southeastern United States. Increased efforts to census and establish population trends of Gulf cetaceans are taking place, and data gathered in the final years of this millennium could help elucidate the trend status of this species in the Gulf of Mexico.

**Remarks.**—It is unlikely (but nonetheless possible) that some purported sightings of short-finned pilot whales in the past were those of the generally less tropical North Atlantic long-finned pilot whales (*G. melas*). No strandings of long-finned pilot whales, however, have been documented for the Gulf.

*Globicephala melas* (Traill, 1809), Long-finned Pilot Whale

**Names.**—The generic name is derived from the Latin *globus,* "a ball," or "globe," and the Greek *kephale,* "head," referring to the rounded forehead or melon. The specific name is derived from *melanus,* or "black." Other common names include blackfish, pothead, pilot whale, and longfin pilot whale. Spanish: calderón común.

**Description.**—The long-finned pilot whale resembles the short-finned pilot whale in all but a few external features. It has a large and stocky body, with males markedly larger than females. Males reach 6.7 m (22 ft.) in length, but females attain only 5.7 m (19 ft.). On average, adult long-finned pilot whales are slightly longer than their more tropical short-finned counterparts. As the name implies, they have extremely long (and thin) flippers with slender and very pointed tips that comprise 18–27 percent of body length. The dorsal fin is about one-third of the way back from the snout tip, and is low and falcate with a wide base. The tailstock is laterally compressed and heavily keeled above and below. It, however, remains more or less uniform in height from a light saddle patch just behind the dorsal fin to just ahead of the flukes. The aft part of the body does not taper gradually like that of most other delphinids. Males have a larger, more bulbous head, thicker dorsal fin, and deeper tailstock than females.

Coloration is very similar to that of the short-finned pilot whale, and it is difficult (if not impossible) to distinguish the two species under normal viewing conditions in nature. Since no strandings of long-finned pilot whales have been documented for the Gulf of Mexico, sightings of live animals are considered short-finned pilot whales unless evidence (e.g., from underwater views of flipper length) indicate otherwise.

**Behavior and Ecology.**—Long-finned pilot whales are highly social, and generally are found in larger pods than short-finned pilot whales. Usual group size ranges from 20 to 100 animals, but groups of 1000 or more are not uncommon in the higher latitudes. They appear to live in stable pods like those of killer whales.

Pilot whales are highly vocal, having both clicks and whistles. They feed largely on squid, often diving more than 500-m (1640 ft.) deep while foraging. While they can be extremely active in the air, they often are encountered rafting at the surface for prolonged periods, apparently at rest.

Long-finned pilot whales appear to be particularly vulnerable to mass strandings. Their tight social structure may be responsible for this; when several animals strand, others invariably follow. This social cohesiveness also has made them vulnerable to humans herding them for mass slaughter, off the north Atlantic Faroe Islands, Newfoundland (in the past), and elsewhere.

**Remarks.**—While no long-finned pilot whales are known to have stranded in the Gulf of Mexico, they occur regularly as far south as Georgia on the eastern U.S. coast. It is not inconceivable that groups might be found in the Gulf at some future date as accidentals or strays.

*Pseudorca crassidens* (Owen, 1846), False Killer Whale

**Names.**—The generic name is from the Greek *pseudo* "false," and *orca,* "a whale." The specific name is derived from the Latin roots *crassus,* "thick," and *dens,* "a tooth." Other common names include thicktooth grampus and pseudorca. Spanish: orca falsa.

**Description.**—These whales are long and slender in body form. The head is narrow and rounded, and the snout overhangs the lower jaw. The dorsal fin is 18- to 41-cm (7–16 in.) high, falcate, and located near the center of the back. Also, the flippers have a hump, or "elbow," at the midpoint of the front margin. Coloration is dark gray to black except for varying white areas around the lips, chin, and belly. Males are typically larger than females, averaging 5.5 m (18 ft.) in length with a maximum of 6.1 m (20 ft.) compared to an average of 4.9 m (16 ft.) with a maximum of 5.5 m (18 ft.) for females. Maximum weight is about 2500 kg (2.5 metric tons) in larger males.

Stranded specimens are easily recognized by the characteristically humped front flippers and the prominent teeth. The teeth are circular in cross section, about 2.5 cm (1 in.) in diameter, slightly curved, and number 7–12 in each side of both jaws (Table 2).

Although much larger, the false killer whale somewhat resembles the pygmy killer whale and melon-headed whale. Distinguishing characters for these three species are given in the accounts for pygmy killer and melon-headed whales.

**Behavior and Ecology.**—False killer whales generally are observed in groups of two to several hundred animals, with both sexes and different age groups usually represented. These whales can emit "whistling" sounds audible to humans and, like all delphinids, are good at echolocation. Squid and fishes make up the diet. They also attack smaller cetaceans, and records exist of attacks against both sperm and humpback whales.



These whales are stranded often, with numerous instances of mass strandings. On January 11, 1970, 150–175 false killer whales beached themselves at three places along the southeastern coast of Florida. The whales appeared about equally divided between the sexes and at least one calf was reported. No cause for the stranding was apparent. Attempts to drag some of the whales to the open ocean met with failure as they returned to the beach to strand again, and all subsequently died. Other mass strandings of this species have been reported in Europe, South Africa, Australia, New Zealand, Ceylon, Zanzibar, Chatham Island, and Argentina, as well as in the Gulf of Mexico.

Their reproductive habits are poorly known. Breeding probably occurs year-round and the gestation period lasts approximately 15 months. Newborns are about 1.5- to 2.1-m (5.0–6.9 ft.) long and weigh at least 80 kg (175 lb.) at birth. Sexual maturity is reached at 3.2 to 3.8 m in length.

**Distribution and Status.**—False killer whales are distributed throughout deep tropical, subtropical, and temperate waters of the world, but not beyond about 50°N or 50°S. In the western North Atlantic, they have been reported off Maryland southward along the mainland coasts of North America, the Gulf of Mexico, and the southeastern Caribbean Sea. There are no worldwide population estimates.

**Status in the Gulf.**—Strandings in the Gulf have been documented 15 times—from Cuba, the Florida Keys, Florida, Louisiana, Texas, and southern Mexico. About one-third of these have been mass strandings, with 3 to about 30 stranded at a time. Most strandings have been in spring or summer, but it is not certain if this represents a seasonal movement of animals.

The pre-GulfCet historical record of live animals sighted at sea included 11 such sightings throughout the northern Gulf in water generally 200 to more than 2000 m (656–6,560 ft.) deep. During GulfCet 1992–97 surveys, another five were sighted by ship and three by airplane. Group sizes ranged from 12 to 63. The abundance estimate is as high as about 1000 animals for the northern Gulf; however, this number must be treated with extreme caution due to paucity of data.



*Feresa attenuata* Gray, 1874, Pygmy Killer Whale

**Names.**—The generic name is probably from *Feres,* a French vernacular name for "dolphin." The specific name is from the Latin *attenuatus,* "thin" or "tapered." Spanish: orca pigmea.

**Description.**—The pygmy killer whale is a small and slender delphinid. Adult males average 2.3 m (7.5 ft.) in length, reaching a maximum of 2.7 m (9 ft.). Females are slightly smaller, averaging about 2.1 m (7 ft.) with a maximum of 2.4 m (8 ft.). Males weigh up to approximately 227 kg (500 lb.) and females, 200 kg (440 lb.).

The head is rounded and there is no prominent beak, although the snout slightly overhangs the tip of the lower jaw. The dorsal fin is falcate, located at the midpoint of the back, and measures 20–50 cm (8–12 in.) in height. The flippers are slightly rounded at the tips.

Coloration is predominantly dark gray to black with white margins about the lips, a white "goatee" on the chin, and a white patch in the anal region. The area between the flippers is gray to white. A narrow dark cape dips slightly below the dorsal fin.

These whales are easily confused with the false killer and melon-headed whales. False killer whales are much larger, lack a white goatee on the chin, and have a bend, or "elbow," in the front margin of the flippers, as opposed to the smoothly rounded flippers of pygmy killer whales. Melon-headed whales also lack the white chin patch, have a narrower head, and have flippers that are pointed at the tips, in contrast to the rounded flippers of false and pygmy killer whales.

In addition to the above characters, stranded specimens can be distinguished by their teeth (Table 2). Pygmy killer whales have 8–11 slender teeth in each side of the upper jaw, and 11–13 in each side of the lower jaw. False killer whales have 10–12 stoutly shaped teeth in each side of both the upper and lower jaws, and the teeth are circular in cross section. Melon-headed whales have 20–25 small teeth in each side of both upper and lower jaws.

**Behavior and Ecology.**—Very little is known about the natural history of this species. Group size is generally 10–50 individuals, although herds of several hundred animals are observed occasionally. In captivity, these whales eat about 9 kg (20 lb.) per day of sardines, squid, sauries, or mackerel. They are quite aggressive and have attacked other delphinids while in captivity. They have been reported to attack other delphinids incidentally caught in tuna nets in the eastern tropical Pacific. Their major food items, however, appear to be squid and fishes. Reproductive habits are not well known, although mating occurs in spring.



**Distribution and Status.**—Pygmy killer whales appear to be restricted to the deep tropical, subtropical, and warm temperate oceans of the world. About twice as many reports of these whales are from tropical waters as elsewhere. In the western North Atlantic, they are known from the southeastern United States as far north as the Carolinas, extreme south Texas, and the West Indies. They have been regarded as rare, but the sparse number of sightings may be due to their generally slow and somewhat "secretive" movements. Seasonal movements remain unknown.

**Status in the Gulf.**—A freshly dead pygmy killer whale was found near Brazos Santiago Pass, South Padre Island, Cameron County, Texas, on January 21, 1969. Detailed observations and measurements of the animal's dentition were made, but little else concerning the stranding was recorded. This specimen was not only the first record for the Gulf of Mexico but also represented the first known occurrence of the pygmy killer whale for the entire western North Atlantic. Thirteen other strandings have been documented, from southern Florida to south Texas, with an apparent peak of strandings in winter.

Pygmy killer whales have been identified from ships four times—once off the south Texas coast in November and three times during spring in the west–central portion of the GulfCet study area, generally in water 500 to 1000 m (1,640–3,280 ft.) deep. Group sizes were 6–30 animals. Abundance in the northern Gulf is estimated at about 400 animals. These whales probably are year-round residents of the Gulf and occur throughout these waters in small numbers.

*Peponocephala electra* (Gray, 1846), Melon-headed Whale

**Names.**—The generic name is from the Greek *pepon,* "gourd," and *kephale,* "head," referring to the asymmetrical, bulbous shape of the head. *Elektra* was a water nymph in Greek mythology. Another common name is many-toothed blackfish. Spanish: calderón pequeño.



**Description.**—Melon-headed whales are small and slender, with adult males reaching 2.7 m (9 ft.) in length and adult females measuring up to 2.6 m (8.5 ft.). Maximum weight is about 275 kg (605 lb.). Newborns are approximately 1 m (3 ft.) in length.

The profile of the melon-headed whale is similar to that of the pygmy killer whale. The head is rounded in profile on the top, flat below and, seen from the top or bottom, forms a distinct triangle between the eyes and the tip of the snout. The dorsal fin is approximately 25-cm (10 in.) tall and is located at the center of the back. The flippers are fairly long with smoothly curved to pointed tips. In comparison, the pygmy killer whale has a completely rounded, bulbous head and flippers with rounded tips.

Coloration is dark gray except on the belly and around the mouth, where a light gray or white patch is present. Although the belly may be very white, it is usually a light shade of gray and is not as distinct as the belly patch of the pygmy killer whale. Also, melon-headed whales may have white pigmentation around the lips resembling that of the pygmy killer whale, and a light stripe from the blowhole to the snout tip. A dark black cape dips strongly below the dorsal fin to a lower position than that observed in the pygmy killer whale and an indistinct dark "mask" often is present on the face.

In the melon-headed whale, both upper and lower jaws have 20–25 sharply pointed teeth per side, about twice the number found in the pygmy killer whale and the false killer whale (Table 2). This is a good distinguishing character for stranded specimens, even for those that are badly decomposed.

**Behavior and Ecology.**—These whales often travel in large groups of 100 to 1500 animals. In the tropical Atlantic, Pacific, and Indian Oceans, they have been reported with Fraser's dolphins (*Lagenodelphis hosei*) and spinner and spotted dolphins of the genus *Stenella.* Melon-headed whales feed mainly on fishes and squid.

Little is known about their reproductive biology. Calving appears to peak during early spring in the low latitudes of both hemispheres, although newborns have been reported in July and August (winter) from the southern hemisphere. The length of the gestation period is not known but is thought to be about 12 months.

**Distribution and Status.**—Melon-headed whales occur worldwide in tropical and subtropical waters to about 40°N and 35°S. Like false killer and pygmy killer whales, they appear to favor warm oceanic waters and rarely stray into the relatively shallow depths over the continental shelf. They are not known to migrate; strandings occur year-round in tropical and subtropical waters.



**Status in the Gulf.**—Melon-headed whales were first recorded from the Gulf of Mexico on the basis of two recent strandings in the western Gulf. On June 22, 1990, a 2.6-m (8.5 ft.) adult male stranded alive on west Matagorda peninsula, Matagorda County, Texas. The animal was pushed back to sea by well-meaning onlookers, but immediately re-stranded. After being taken to deeper water for a second attempted release, it swam off; in late afternoon, however, it was found dead on a nearby beach. This specimen represented the first documented occurrence of the species in the Gulf and clarified the worldwide distribution of these whales in tropical and warm temperate waters. Interestingly, a second melon-headed whale, a 2.5-m (8.4 ft.) male, was discovered on June 14, 1991, in Cameron Parish, Louisiana, almost one year later in a state of "moderate" decomposition.

In late summer of 1998, a young male stranded alive south of Corpus Christi, Texas; the probable mother was found 15 km (9.3 mi.) from the calf, but died after a few hours.

During the 1992–94 GulfCet study, melon-headed whales were sighted 10 times by ship and 4 times by air, in 200- to 2000-m (656–6,560-ft.) deep waters throughout the northwestern Gulf study area. During the 1995–97 GulfCet II study, only one group of 125 was sighted in the northwestern Gulf during spring. Five additional sightings of either pygmy killer whales or melon-headed whales were made by air, but could not be assigned to either species with certainty. Of the 15 confirmed sightings, 9 occurred in the mid-western portion of the study area, 1 off south Texas, and 5 south of the Mississippi Delta. Group sizes ranged from 30 to 400 animals, and abundance in that northwestern area has been estimated about 2000 animals. Melon-headed whales were sighted with Fraser's dolphins during four Fraser's dolphin group sightings of the GulfCet surveys. Thus, it may be that melon-headed whales occur in only a few groups but in large numbers in at least the northern part of the Gulf.



*Steno bredanensis* (Lesson, 1828), Rough-toothed Dolphin

**Names.**—The generic name is from the Greek *stenos,* "narrow." The specific name is in honor of the artist (Van Breda) who first portrayed this species. Another common name is steno. Spanish: esteno.

**Description.**—This is a fairly small delphinid with a maximum length of 2.7 m (9 ft.) for males and 2.3 m (7.5 ft.) for females. Weight averages 130 kg (285 lb.) with a maximum of about 158 kg (350 lb.).

Unlike many other dolphins in which the beak is sharply delineated from the melon, rough-toothed dolphins have a prominent beak that slopes smoothly into the forehead, giving the head a conical shape and emphasizing the sleek, symmetrical lines of the body. The flippers, set far back on the body, and the large dorsal fin are smoothly curved and pointed. The tailstock is keeled, especially in adult males.

Dorsally, the coloration is dark gray to black and may have a purplish cast. The belly is white and tinted with pink, and the sides of the beak, throat, and chin are white. Yellowish white spots occur on the sides, and these dolphins are often scarred. Rough-toothed dolphins often are confused with the bottlenose dolphin (*Tursiops truncatus*), but the white pigmentation of the sides of the beak and conical shape of the head set it apart. Their tooth morphology is also unique (Table 2). The 20–27 teeth in each side of both jaws have fine lateral ridges extending to the crown, giving a rough, sandpaperlike texture to the tooth surface. The combination of conical head shape and characteristic, rough-textured teeth readily distinguishes stranded specimens.

**Behavior and Ecology.**—Little is known about their natural history. They occasionally may travel in large groups of 100 or more but smaller schools of 10 to 20 animals are normal. They often ride the bow waves of boats, and will at times even stay with very slowly moving vessels. When traveling rapidly, they often surface with a peculiar head-up motion, with the neck area skimming along the surface of the water. They are considered one of the most intelligent of cetaceans and are easily trained to perform aquarium antics. Their diet consists of octopus, squid, and fishes, and they are known to take large fish such as mahi-mahi (or dorado, *Coryphaena hippurus*) in tropical waters.

Virtually nothing is known about their reproductive habits. In Hawaii, the mating of a captive female rough-toothed dolphin with a male bottlenose dolphin produced a hybrid calf that survived five years. The fertility of this hybrid was not determined by the time of its death.



**Distribution and Status.**—Rough-toothed dolphins are distributed in tropical and warm temperate waters of the world. Records from the western North Atlantic are from Virginia down to Florida, the Gulf of Mexico, the West Indies, and the northeastern coast of South America. These are generally offshore, deepwater dolphins. There are no reliable population estimates and seasonal migratory tendencies are not known.

**Status in the Gulf.**—The rough-toothed dolphin has stranded in 10 different areas of the northern Gulf of Mexico, from south Texas to the keys of Florida. One of us (DS) reported the first rough-toothed dolphin known from Texas waters, found by two small boys in late June of 1969; it was beached on the northern shore of San Luis Pass at the western end of Galveston Island. An excellent photograph of the specimen, unquestionably *Steno bredanensis,* was included with the account.

Two more recent strandings of rough-toothed dolphins in the Gulf of Mexico were reported to the Southeastern United States Stranding Network. On September 6, 1985, a 254-cm (100-in.) male stranded alive on Bolivar peninsula, Galveston County, Texas. This animal was transported to Sea-Arama in Galveston where it died on September 24. On October 18, 1987, three rough-toothed dolphins stranded alive on Key Largo, Dade County, Florida. These animals were all males, measuring and weighing 234 cm and 113 kg (249 lbs.), 249 cm and 110.5 kg (243 lbs.), and 239 cm and 102 kg (224 lbs.).

Rough-toothed dolphins have mass stranded on three occasions in the Gulf of Mexico. Fifteen stranded at Cedar Key, Levy County, Florida, on November 24, 1974, and there was a live stranding of approximately 30 animals near Key West, Monroe County, Florida.

The best documented mass stranding in the Gulf of Mexico occurred when 16 animals came ashore on the Gulf coast of Florida on the night of May 29, 1961. Due to the location of the carcasses and the lack of visible signs of struggle, the dolphins may have stranded farther out during low tide and drifted closer to shore at high tide. Six were males and four were females; the remainder could not be sexed. None of the females were pregnant. The stomachs of four dolphins were empty, but the remainder had ingested blanket octopus (*Tremoctopus violaceus*) and the alga *Sargassum ilipefundula.* Roundworms were found in the stomach of one individual.

Stranding and sighting records in the Gulf of Mexico are from all seasons, and it is likely that rough-toothed dolphins occur there year-around. During the GulfCet surveys, they were also sighted in all seasons. Group sizes averaged about 10 animals, with a range of 2 to 48 individuals. Abundance in the northern Gulf of Mexico is estimated at about 500 animals but could be considerably higher.

*Grampus griseus* (Cuvier, 1812), Risso's Dolphin

**Names.**—*Grampus* is New Latin for "whale." The specific name is from the Latin *griseus,* "gray." Another common name is grampus. Spanish: delfín de Risso.

**Description.**—This is a medium-sized dolphin, averaging over 3 m (10 ft.) in length; adults have a maximum length of about 3.8 m (12.5 ft.). Average weight is around 300 kg (660 lb.) but very large individuals may weigh up to 500 kg (1100 lb.).

Risso's dolphin is stocky and robust in body form, although the tailstock is quite slender. The characteristically shaped head is blunt and without a beak, in profile presenting a "blocky" or squarish aspect. Also, the forehead, or melon area, is divided medially by a shallow crease. The dorsal fin is relatively tall, up to 51 cm (20 in.), and falcate. The flippers are long and pointed, and the broad tail flukes are notched medially.

Coloration is dark gray dorsally with lighter gray or white patches ventrally. In older individuals, the face and area just forward of the dorsal fin fade to a light gray, contrasting with the darker pigmentation of the side, back, extremities, and tailstock. Newborns are a uniformly light gray that gradually darkens as the animals age. Risso's dolphins are generally heavily scarred, perhaps by parasites and squid bites, as well as by wounds inflicted in intraspecific fighting. Such scarring appears as numerous light-colored scratches, streaks, or thin lines that stand out against the darker ground coloration. Due to heavy scarring and a general loss of dark pigmentation in adults, older Risso's dolphins are usually very light in color, especially around the head; coloration can be quite variable within populations.



The blunt, creased head and extensive scarring are good field identification characters. These dolphins are sometimes observed riding bow waves of vessels and spyhopping. Viewed from the air, contrast between the pale and dark portions of the body is enhanced.

Stranded specimens are distinguished by the squarish head shape, medial head crease, and two to seven teeth in each side of the lower jaw which are absent from the upper jaw (Table 2). Some or all teeth may be worn down or completely missing in older adults.

**Behavior and Ecology.**—These dolphins have been observed in groups as large as 4000 animals, but smaller ones of 3 to 30 are more common. They often swim slowly and with few surface displays, but are sometimes seen porpoising (leaping well above the water to breathe) and breaching. They often associate with other cetacean species, and hybrids between Risso's and bottlenose dolphins have been recorded in captivity and in nature. They feed primarily on squid, but also eat fishes and crustaceans.

Their reproductive biology is poorly known, but there appears to be a summer calving peak in the North Atlantic. A famous Risso's dolphin, known as "Pelorus Jack," was sighted in a New Zealand harbor for more than twenty years, and they probably can live at least twice that long.

**Distribution and Status.**—This is an offshore, deepwater species that is distributed worldwide in warm temperate and tropical waters. In the western North Atlantic, it is found from Newfoundland southward to the Gulf of Mexico, throughout the Caribbean, and bridging the equator. It was considered uncommon in the past, but this apparent rarity is due almost certainly to its oceanic range, which is often outside normal shipping lanes.

Migratory movements are not well documented, but seasonal increases in population densities may occur in some areas. For example, there is evidence of seasonal movements in the eastern North Pacific that probably correlate with prey movements associated with water temperatures.

**Status in the Gulf.**—There are sixteen records of stranding sites for the Gulf, concentrated in Texas and Florida (including four in the Keys), with one off the northwestern shore of Cuba.

Aerial surveys conducted in the Gulf of Mexico and Atlantic waters of Florida in 1980 and 1981 located approximately 274 Risso's dolphins on 12 different occasions. Seven sightings (67 animals) occurred in Atlantic waters east of Florida, four sightings (191–98 animals) were made off western and southwestern Florida in the Gulf of Mexico, and one sighting of a group of nine animals occurred off the south Texas coast. The Texas sighting was made on November 1, 1980, and represented the first record of Risso's dolphin in the western Gulf.

During the 1980–81 surveys, pelagic sightings occurred in waters 200 to 1530 m (656–5,020 ft.) deep, with most of the dolphins appearing to prefer deep continental slope waters. In recent years (late 1980s–94), however, over 60 sightings of groups have been made, most commonly over or near the 200-m depth contour just south of the Mississippi Delta. It is not known if this represents a more recent incursion of Risso's dolphins to these shallower waters or simply that they were missed during earlier studies. This dolphin is probably a year-round resident of the Gulf, and is common in some continental slope areas.

During the GulfCet surveys, these dolphins were sighted in all seasons a total of 161 times. Group sizes ranged from 1 to 78 individuals. Abundance is estimated as high as about 4000 animals in the northern Gulf.

*Tursiops truncatus* (Montagu, 1821), Bottlenose Dolphin

**Names.**—The generic name is from the Latin *tursio,* "a porpoise," and the Greek *ops,* "the face." The specific name is derived from the Latin root *truncase,* "cut off," or "short." Literally translated as "short porpoise face," the scientific name refers to the short beak of this species. Spanish: tursion or delfín nariz de botella.

**Description.**—Bottlenose dolphins are medium-sized dolphins that average 1.9–3.8 m (6.0–12.5 ft.) in length, with much variation among populations. Average weight is approximately 200 kg (450 lb.). Maximum size is about 4.3 m (14 ft.) in length and 650 kg (1430 lb.) in weight. Specimens from the Gulf of Mexico usually measure 2.7 m (9 ft.) or less in length.

The head is distinctive in abruptly tapering to a relatively short and robust snout, with the lower jaw slightly protruding beyond the upper. A shallow crease surrounds the base of the snout and sets it off from the melon. The dorsal fin is falcate, usually tall (also variable by population), and positioned at the middle of the back. Coloration is gray dorsally but grades to a lighter gray on the sides and whitish gray ventrally. There is incredible individual and geographic variation in color from very dark to light gray, with variable dorsal capes and saddlelike light patches.



*Tursiops* often will approach ships to ride the bow wave. These dolphins are distinguished at sea by their short beaks and even coloration that usually lacks spots or mottling. Stranded specimens are also distinguished by the characteristic snout. The teeth number 18–26 per side in both upper and lower jaws.

**Behavior and Ecology.**—Bottlenose dolphins may be observed in groups numbering up to several hundred, but smaller social units of 2 to 15 members are more common. Group size is affected by habitat structure and tends to increase with increased water depth, or "openness" of habitat. Group members interact closely and are highly cooperative in feeding, protective, and nursing activities. Possible "scouting" behavior has been documented, whereby individuals investigate obstacles and then lead the group past the barriers. Examples of adult dolphins helping other injured adults and adults supporting injured or even dead infant dolphins, are also well known. In captivity, dominance hierarchies serve to structure bottlenose groups, whereas in the wild, dominance probably is expressed by sex and age segregation of groups.

*Tursiops* makes numerous sounds and is proficient at echolocation. As early as 1940, noises made by captive dolphins were believed to have communicative value. These included a "snapping noise made with the jaws" used to intimidate subordinates, and "whistling" and "barking" sounds made underwater. These dolphins are now known to have a huge repertoire of sounds, with slight differences among animals from different populations.

Bottlenose dolphins eat a wide variety of food items depending on what is available and abundant at a given time. In Texas waters, they eat fishes, including young tarpon, small sharks, speckled trout, pike, rays, mullet, catfish, and shrimp. In the northern Gulf, anchovies, menhaden, mullet, minnows, shrimp, and eel have been reported in their diet. *Tursiops* eat approximately 9 to 18 kg (20–40 lb.) of fish per day.

Seven recurrent feeding behaviors have been described for the northern Gulf of Mexico: (1) foraging behind working shrimp boats and eating organisms disturbed by the nets; (2) feeding on trash fishes dumped from the decks of shrimp boats; (3) feeding on fishes attracted to nonworking shrimpers; (4) herding schools of fishes by encircling and charging the school or feeding on the stragglers; (5) sweeping schools of small baitfishes into shallow water (ahead of a line of dolphins charging into the school or feeding on the stragglers); (6) crowding small fishes onto shoals or mudbanks at the base of grass flats (driving fish completely out of the water and then sliding onto banks to retrieve them); and (7) individual feeding. Feeding methods, such as feeding behind shrimp trawlers and lunging onto mudbanks, may be local traditions that probably were learned by succeeding generations. In coastal Texas waters, feeding near shrimp trawlers is especially widespread, and the dolphins feed on fishes and invertebrates discarded by the trawlers as well as directly on shrimp.

Mating and calving primarily occur from February through May,

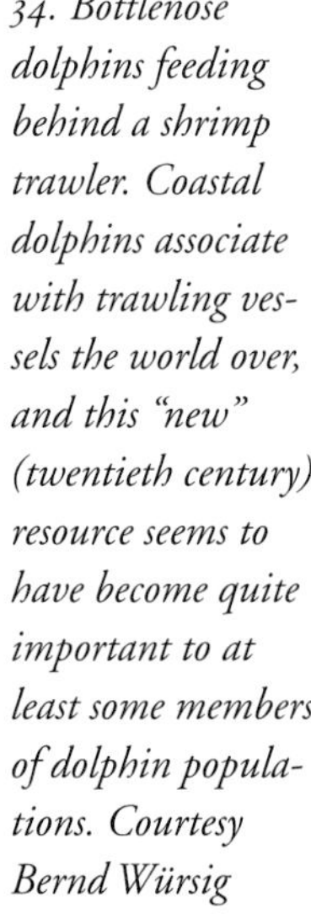

*34. Bottlenose dolphins feeding behind a shrimp trawler. Coastal dolphins associate with trawling vessels the world over, and this "new" (twentieth century) resource seems to have become quite important to at least some members of dolphin populations. Courtesy Bernd Würsig*

*35. Bottlenose dolphins surfacing in a welter of foam during social/sexual activity. Although the casual view indicates that the dolphins are having fun, there is often quite a bit of aggression, biting, and tooth-raking in these interactions. Courtesy Thomas Henningsen*

but additional breeding may also take place during summer and fall. Females give birth to a single calf only once every 2–3 years, after a 12-month gestation period. A lifetime total of eight calves is believed typical for females.

Newborn calves are about 0.9-m (3 ft.) long at birth, weigh 9–14 kg (20–30 lb.), and nurse up to 2.5 years following birth. Males mature at 10 to 13 years of age when they are approximately 2.4-m (8 ft.) long, and females mature at 5 to 12 years at approximately 2.3 m (7.5 ft.) in length.

**Distribution and Status.**—Bottlenose dolphins are distributed worldwide in tropical and temperate waters. In the western North Atlantic, they occur as far north as Nova Scotia but are most common in coastal waters from New England to Florida, the Gulf of Mexico, the Caribbean, and southward to Venezuela and Brazil. They are primarily a nearshore species and especially common near passes connecting bays to the open ocean. They also are seen in lagoons, rivers, and in open ocean areas. Pelagic or open ocean bottlenose dolphins are often of different size or color than their closest conspecifics near shore, suggesting little or no interbreeding between offshore and inshore groups.

Some temperate populations may migrate to warmer latitudes in fall, but most do not appear to undertake extended migrations regularly. Local diel movements are common and well documented;

they appear to be related to tidal flow and time of day but also are due in large part to diel movements of fishes.

**Status in the Gulf.**—Bottlenose dolphins are the most widespread and common cetaceans of the coastal Gulf of Mexico, frequently seen in bays, ship channels, and even estuaries. Two distinct forms occur in the Gulf, an inshore group that inhabits shallow lagoons, bays, and inlets and an "oceanic" population that remains in deeper, offshore waters on the continental shelf. The shelf form lives sympatrically with the Atlantic spotted dolphin, which can be confused with the bottlenose (especially young of both species). Interaction between the two populations of bottlenose dolphins is thought to be minimal, and recent genetic evidence points to little or no mating between the two types.



Gulf populations were exploited by sport hunting in Texas during the first half of this century and by a small fishery that was located in eastern Florida near the turn of the century. All cetaceans, including bottlenose dolphins, are now protected from hunting but are deleteriously affected by other human activities. In the Gulf, these include petroleum resources development, heavy boating traffic, and the pollution of Gulf waters; the cumulative effects of these factors on dolphins, however, are difficult to determine. Bottlenose dolphins in Texas waters have been observed swimming through heavy oil spills, and appeared to suffer no immediate ill effects. These nearshore dolphins seem to be well adapted to human activities, but are affected by pollutants and would make a good "indicator species" for the Gulf; increased stranding of sick and injured animals would indicate excessive pollution of Gulf waters.

Although bottlenose dolphins are numerous in Gulf of Mexico waters, exact population numbers are not known. It is not clear whether inshore bay and nearshore coastal dolphins constitute one population or several, and the extent of genetic interplay between nearshore and deeper water dolphins (although it is apparently little) is not known. The best estimate for the number of bottlenose dolphins in the northern Gulf of Mexico appears to be around 78,000 individuals—about 5000 in bays, 18,000 along the coast, and 55,000 in deeper continental slope waters.

During the GulfCet surveys, bottlenose dolphins were almost exclusively sighted at depths of less than 1000 m (3,280 ft.), indi-

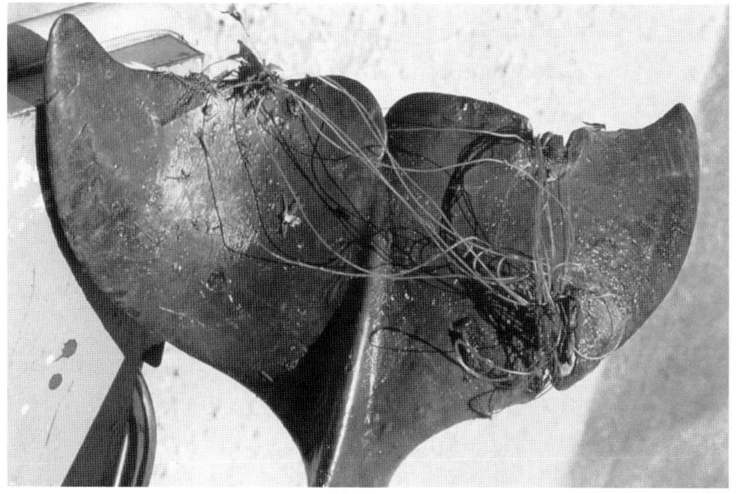

*36. The tail, or flukes, of a hapless bottlenose dolphin ensnared by a fishing net along the Texas shoreline. Accidental entanglements and deaths due to pollution or other forms of anthropogenic habitat changes are all too common along the shores of the Gulf of Mexico. Courtesy Thomas Henningsen*

cating that they do not occur in the deeper, central parts of the Gulf. Abundance in the northwestern shelf area was estimated at about 1500 to 4000 animals, but it would not be surprising to see higher estimates following future work in this area.

**Remarks.**—Bottlenose dolphins are one of the most familiar of cetaceans. They are a ubiquitous and popular attraction at marine aquariums, and have been successfully maintained in captivity since the turn of the century. A great deal of information has accrued on captive bottlenoses and free-range populations. Nevertheless,

*37. Bottlenose dolphins surfing the bow pressure wave of a large vessel. This is a common form of assisted travel—and is probably also fun—for dolphins of the bays, channels, and major coastal waterways of the Gulf of Mexico. Courtesy Thomas Henningsen*

the taxonomy of this genus is in flux, and it is likely that several distinct species will be recognized due to recent and ongoing genetic studies.

*Stenella attenuata* (Gray, 1846), Pantropical Spotted Dolphin



**Names.**—The generic name is from the Greek *Stenos,* "narrow," and refers to the long and slender beak characteristic of this genus. The specific name is from the Latin *attenuat,* "thin" or "narrow." They are often called spotters. Spanish: estenela moteada.

**Description.**—This is a small dolphin, ranging from 1.6 to 2.5 m (5–8 ft.) in length and averaging 100 kg (220 lb.) in weight, with adult males being slightly larger than females. Maximum size is about 2.7 m (9 ft.) and approximately 136 kg (300 lb.). Offshore animals tend to be slightly smaller than those from coastal populations.

This dolphin is somewhat robust in body form and has a relatively long beak. The flippers are long and pointed, the dorsal fin is tall, narrow, and prominently curved backward, and the tailstock is keeled ventrally, especially in adult males. Coloration is highly variable, especially between age classes. Adults are generally dark gray dorsally and a medium gray ventrally, with numerous white spots speckling the dorsal body. The eyes are set within larger black spots connected across the beak by a black band, or "bridle," and a dark line extends from the flipper to the lower jaw. Other markings include a dark and unspotted dorsal fin, flippers, and tail flukes, and a black beak with a white tip and lips. Newborns are not spotted.

The pantropical spotted dolphin can be confused with the Atlantic spotted dolphin, *Stenella frontalis. S. frontalis* is stockier, has dark spots ventrally (like young *S. attenuata*), and lacks the dark bridle line. Stranded specimens are difficult to distinguish if the body markings have faded. The two spotted dolphins have fewer total teeth than other species of *Stenella,* but the tooth counts are similar for *S. attenuata* and *S. frontalis* (Table 2). *S. frontalis* has approximately 70 vertebrae while *S. attenuata* has about 80; the two species do not overlap in this character.

**Behavior and Ecology.**—This dolphin is usually seen in groups of 5 to 30 animals for coastal forms, but large herds of well over 1000 are common in oceanic waters. It feeds at or near the surface on fishes, including horse mackerel and flying fish, squid, and shrimp; it often associates with spinner dolphins (*S. longirostris*) in the tropical Pacific.

Growth and reproduction have been studied in *S. attenuata* killed by tuna purse seiners in the eastern tropical Pacific. Gestation period is 11.5 months. At birth, calves average 0.8 m (2.7 ft.) in length, and at one year average 1.4 m (4.5 ft.). Males attain sexual maturity at approximately six years of age, and females at five years. The average calving interval is 26 months, which consists of 11.5 months of gestation, 11.2 months of lactation, and 3.3 months of resting and/or estrous. These details, taken from thousands of dead animals, may not be exactly the same for populations in the Gulf of Mexico.



**Distribution and Status.**—The pantropical spotted dolphin is distributed throughout the tropical and subtropical waters of the world. In the western North Atlantic, it is found from North Carolina to the Antilles, West Indies, and down to the equator.

In the Pacific, they are killed incidentally in the course of purse seining for tuna because nets are set around the dolphins to catch the tuna below them. In 1970, approximately 400,000 dolphins were killed by U.S. vessels alone, but that number was down to 15,000–20,000 animals by 1978. By the late 1980s, however, the kill had climbed again to more than 100,000 dolphins per year. It now has declined again to only several thousand per year, due in large part to better dolphin-exclusion techniques. Presently, few U.S. tuna vessels are setting nets around dolphins, and the kill comes mainly from several Latin American countries.

**Status in the Gulf.**—Although pantropical spotted dolphins were once thought to be uncommon in the Gulf of Mexico, they are now known to be the most common cetacean in deep Gulf waters. It is probable that they were often misidentified in the past as Atlantic spotted dolphins (*S. frontalis*). At least 21 stranding records exist, from southern and northern Texas, the coasts of Alabama and Mississippi, Florida, the Florida Keys, and northwestern Cuba. Approximately 300 sightings were made of pantropical "spotter" groups in waters deeper than 100 m between 1989 and 1997. These occurred mainly in the north–central Gulf between the area south of the Mississippi Delta and the 200-m drop-off zone west of Florida. They also occur widely outside this area. From the GulfCet surveys in the northern area of the Gulf, population estimates of almost 60,000 animals were derived, many occurring well beyond the 100-m (328 ft.) depth contour and into waters 2000 m (6,560 ft.) deep.



*Stenella frontalis* (Cuvier, 1829), Atlantic Spotted Dolphin

**Names.**—The specific name is derived from the Latin roots *front,* "the forehead" or "brow," and *alis,* "pertaining to." Spanish: delfín pintado.

**Description.**—This is a small, stocky dolphin that reaches a maximum length of 2.4 m (8 ft.), although 2.1 m (7 ft.) is typical. Average weight is about 109 kg (240 lb.) but may reach 145 kg (320 lb.).

These dolphins are dark gray dorsally, paler gray or white below and, as their name implies, extensively speckled. Dorsally, the spots are pale against the dark base coloration but ventrally, darker and contrast with the lighter gray of the belly. Newborn dolphins are unspotted but begin to acquire spots by their first year. The characteristic spotting of this species becomes more extensive with age. In some offshore populations, however, adults may have little spotting.

The flippers, dorsal fin, and flukes are dark and unspotted, and a pale area, or "blaze," is present on the flank below the dorsal fin. The combination of dorsal spotting and spinal blaze is unique to *S. frontalis.* The upper jaw is white-tipped and the lower jaw is whitish.

These dolphins superficially resemble the pantropical spotted dolphin, *Stenella attenuata,* but the two differ in coloration, with adult *S. attenuata* having gray bellies and adult *S. frontalis,* white bellies. Stranded specimens are difficult to distinguish if the coloration has faded, because the two species overlap in all skull measurements and have similar tooth counts (Table 2). They, however, can be distinguished by the number of vertebrae, which does not overlap; *S. frontalis* has approximately 70 vertebrae and *S. attenuata* has about 80.

Young Atlantic spotted dolphins also may be easily confused with bottlenose dolphins, *Tursiops truncatus,* as they lack the characteristic spotting of adults. Young *S. frontalis,* however, have a more slender head and a longer beak than *T. truncatus,* and the top of the snout is often whitish, rather than gray like *T. truncatus.*

Recent genetic information indicates that *S. frontalis* is closely related to the genus *Tursiops*, and a reclassification may be in order if this relationship is validated upon further scientific scrutiny.

**Behavior and Ecology.**—Atlantic spotted dolphins are observed in herds of up to 50 animals, but smaller groups of 6 to 10 individuals are more common. They feed on small fishes, including herring, anchovies, flounder, mojarralike fishes, carangid fishes, and squid. They appear to cooperate in the herding of schooling fish at the surface, and studies in the Bahamas indicate that they have a fluid social structure, with day-to-day changes in group affiliations and group size. They mate and calve in summer, although animals in the tropics probably have a very protracted breeding season.



**Distribution and Status.**—These dolphins are common offshore inhabitants of tropical and warm temperate waters of the Atlantic Ocean, but are not known outside of the Atlantic. There are numerous records from the eastern coast of the United States (as far north as New England), the Gulf of Mexico, the Caribbean down to southern Brazil, as well as scattered records from the central Atlantic and western Africa. Populations are usually found more than 8 to 20 km (5–12 mi.) offshore, but some animals may move closer to shore during spring and summer. Seasonal movements, however, are not well documented.

**Status in the Gulf.**—This is a common offshore dolphin of the Gulf of Mexico. The species is normally observed from about the 20- to the 200-m (66–656-ft.) depth curves, with a few scattered records out to 1000 m (3,280 ft.). It occurs extensively off the Mexican Campeche Bank north and west of the Yucatan peninsula.

These dolphins may move inshore during the late spring and summer months, or even in winter months when the water is unusually warm. Such movement may be related to the spring arrival of carangid fishes known as "hardtails" (genus *Caranx*). Although seasonal nearshore–offshore movements are not well substantiated, it appears that such movements may be influenced by prey availability or water temperature in the Gulf of Mexico. Interestingly, this abundant continental shelf species is marked by very few strandings.

In the northern GulfCet survey area, group sizes averaged 18–23 animals, with a range of 3 to 75. The population in the western portion of this area was estimated as about 500 to 2000. Atlantic spotted dolphins occur in that area between the 100- and 2000-m (328–6560-ft.) depth contours, and are almost exclusively concentrated around the 100-m (328-ft.) isobath.

**Remarks.**—Gulf populations of this species were previously known as *S. plagiodon,* the scientific name commonly used in the historical literature.



*Stenella longirostris* (Gray, 1828), Spinner Dolphin

**Names.**—The specific name is from the Latin *longus,* "long," and *rostrum,* "a beak." Spanish: estenela giradora.

**Description.**—This is a small dolphin that averages only 1.8 m (6 ft.) in length and 75 kg (165 lb.) in weight. Maximum size is about 2.1 m (7 ft.) in length and 95 kg (210 lb.) in weight, with males slightly longer than females.

This dolphin is slender in form and has an extremely long and slender beak. The flippers are large and pointed, and the dorsal fin is relatively tall and only slightly curved backward, or triangular. Coloration is dark gray dorsally, fading to lighter gray on the sides, and the belly is white. A dark stripe extends from the flipper to the eye. The dorsal fin, flippers, and tail flukes are uniformly dark. Finally, the beak is black above and white below, and the tip is dark. There are many geographic forms of the spinner dolphin, with coloration varying widely from highly distinct to muted capes and lines. The difference between males and females also varies, and at least one form in the eastern Pacific shows strong sexual dimorphism, with adult males having high, erect dorsal fins and large post-anal keels. In the Atlantic, this dolphin has often been confused with the Clymene dolphin (*S. clymene*), but the latter is shorter and has distinctively different markings. A dwarf form of spinner dolphin exists in the Gulf of Thailand.

The spinner dolphin derives its name from a habit of leaping high from the water and rapidly spinning up to six times around the long axis of the body, either pointed straight up out of the water, or at an angle. Such distinguishing leaps occur during high levels of social activity, and appear to be an outgrowth of sociality, or "excitement."

These dolphins do not strand often, but stranded specimens are easily distinguished by the number of teeth. With an average of 45 to 62 teeth per each jaw, spinner dolphins have more teeth than the average of any other oceanic cetacean, but the numbers of teeth overlap with other species of *Stenella* and *Delphinus* (Table 2).

**Behavior and Ecology.**—Spinner dolphins usually occur in groups of 30 to several hundred individuals, and even the low thousands. Around deepwater islands and atolls, they exhibit a fission–fusion society with hundreds feeding together at night, but only dozens in shallow nearshore waters during daytime. They feed on mesopelagic fishes and squid at night.



Adult females give birth to a single calf at two-year intervals. Parturition usually takes place in early summer but can occur in any season. The period of gestation is approximately 11 months, and calves are about 0.75-m (2.5 ft.) long at birth. They reach sexual maturity at 1.5 to 1.8 m (5–6 ft.) in length.

**Distribution and Status.**—Spinner dolphins are offshore, deepwater dolphins that occur worldwide in tropical and warm temperate waters. They also occur close to tropical islands and atolls with adjacent deep water. In these areas, they typically rest and socialize in the shallows during the day and feed in deep water at night. Their range is almost the same as that of pantropical spotted dolphins. In the western North Atlantic, they are known from South Carolina to Florida, the Caribbean, the Gulf of Mexico, and the West Indies southward to Venezuela.

Spinner dolphins have been reduced greatly in the eastern tropical Pacific by incidental killing in tuna purse seines. In the 1970s and 1980s, populations appear to have declined from about 2 million down to about 400,000 animals (an 80 percent reduction).

**Status in the Gulf.**—This dolphin has mass stranded twice in the Gulf of Mexico. Approximately 36 spinner dolphins stranded on Dog Island, Florida, in September of 1961. The stranding site was of gently sloping sand, and it is likely that the animals beached at low tide. From July 13 through 16, 1976, about 50 spinner dolphins stranded at several points near Sarasota, Florida. The dolphins initially beached on the evening of July 13 during an extremely low tide. Several of the animals were successfully returned to the sea; however, others merely stranded again and at least 28 died.

During the Fritts surveys of the 1980s, sightings of this species were purportedly made in waters less than 200 m (656 ft.) deep west of Florida and on the Mexican Campeche Banks; however, it is now thought that these animals were probably misidentified.

During the GulfCet surveys, spinner dolphins were sighted in every season except fall with group sizes ranging from 9 to 750 animals. At least 15,000 spinner dolphins are believed to occur in the northern Gulf, generally in waters much deeper than 100 m (328 ft.). Almost all sightings have been east and southeast of the Mississippi Delta.



*Stenella clymene* (Gray, 1850), Clymene Dolphin

**Names.**—The specific name is from the Greek *Clymene,* daughter of Tethys and Oceanus. A recent common name is short-snouted spinner dolphin. Spanish: delfín clymene.

**Description.**—This is a small dolphin that averages only 1.8 m (6 ft.) in length and 75 kg (165 lb.) in weight. Coloration is black dorsally, light gray on the sides and tailstock, and white ventrally. Dorsally, the surface of the head is marked by a black band that extends from the tip of the beak to the melon, then continues to the blowhole as a lighter, gray band. The remainder of the beak is gray above with white below, and the lips are black. Other markings include dark flippers and tail flukes, and a gray dorsal fin bordered with dark margins. The most distinctive character of this species, however, is the two-tone, beak-to-blowhole stripe and a "mustache" marking.

This dolphin closely resembles the spinner dolphin (*S. longirostris*), but is slightly smaller and more robust in body form and has a shorter beak and fewer teeth. The teeth number 43–58 in each side of both jaws and average 200 total (Table 2). Skull morphology indicates that this species may be closely allied to the striped dolphin (*S. coeruleoalba*).

**Behavior and Ecology.**—This dolphin was not fully described as a distinct species until 1981, and not a great deal more has been learned about its biology. It has been observed at sea only in deep water (250–5000 m /820–16,400 ft. or deeper). It eats small fishes and squid, and appears to be a mesopelagic (mid-water) and night feeder. Squid remains found in the stomachs of some specimens are of species that characteristically live at intermediate ocean depths and surface at night.

Clymene dolphins have been observed leaping and spinning out of water, but their movements are not as high or complex as those of *S. longirostris.* They have been observed with other species of delphinids. They are usually quick and appear agile as they surface, and frequently ride bow waves.

**Distribution and Status.**—The Clymene dolphin is distributed in the tropical and subtropical waters of the Atlantic Ocean, but has not been reported from tropical regions of the Indian or Pacific Oceans. This is unusual for a tropically distributed cetacean, and it has been speculated that *S. clymene* may have evolved in the tropical Atlantic during the Pleistocene era. In the western Atlantic, this dolphin is known from New Jersey to Florida, the Caribbean, the Gulf of Mexico, and down to Venezuela and Brazil.



In the past, *S. clymene* was regarded as a rare cetacean, but this is certainly a result of taxonomic confusion with *S. longirostris,* from which it was recently differentiated.

**Status in the Gulf.**—There are about 70 verified records from the Gulf of Mexico, indicating that the species is not rare. During the GulfCet surveys, these dolphins were sighted most numerously in a central portion of the study area well past the 100-m (328-ft.) isobath, with fewer sightings east of a north–south line drawn at the Mississippi Delta and west of one drawn at Galveston Bay. Group sizes ranged from 2 to 200 animals. About 2000 animals have been estimated to occur in the nearshore east, and as many as 10,000 in oceanic waters of the northern Gulf. They were more widely distributed in the western oceanic Gulf during spring and in the northeastern Gulf during summer and winter.

### *Stenella coeruleoalba* (Meyen, 1833), Striped Dolphin

**Names.**—The specific name is derived from the Latin roots *caeruleus,* "sky-blue," and *albus,* "white." Other common names include the blue-white dolphin or euphrosyne dolphin. Spanish: estenela listada.

**Description.**—This dolphin is slender in form and slightly larger than other members of the genus *Stenella.* Striped dolphins reach average lengths of about 2.4 m (8 ft.) with a maximum of approximately 3 m (10 ft.). Weight averages 100 kg (220 lb.) but may reach 129 kg (285 lb.). They appear more robust than spinner and pantropical spotted dolphins.

Striped dolphins are strikingly colored in shades of gray or brown. The dorsal coloration is dark but fades into lighter colored sides and a white ventrum. A distinctive black stripe extends along the side from the eye to the anus and another stripe extends from the eye to the flipper. Branching from the side stripe above the flipper, a black process extends down to the belly. Also, a process of dark dorsal coloration extends from behind the fin forward and into the lighter flank. Other characters include a tall and strongly curved dorsal fin, and a uniformly dark beak of moderate length.

Striped dolphins may be confused with other stenellids and common dolphins (*Delphinus* spp.), but these other species lack the distinctive pattern of striping characteristic of *S. coeruleoalba.* Stranded striped dolphins are distinguished by their coloration and their teeth. Striped dolphins have 40–55 teeth in each side of both jaws, and average 200 total teeth (Table 2). Only *S. longirostris* has more teeth (224 total) than the striped dolphin.

**Behavior and Ecology.**—Striped dolphins have been observed in herds of several hundred to several thousand animals, with such groups apparently segregated by age and sex at times. Diet consists of small mesopelagic squid and fishes, especially of the lanternfish (myctophid) group.

Calves are approximately 1-m (3 ft.) long at birth. Sexual maturity is reached at about nine years of age for males and seven years for females. Adult females bear young once every 3 years and the gestation period is about 12 months.

**Distribution and Status.**—The striped dolphin is distributed worldwide in tropical and warm temperate waters, at latitudes between 50°N and 40°S. In the western North Atlantic, it is known from Nova Scotia southward to the Caribbean, the Gulf of Mexico, and Brazil. It is a deepwater species and comes near shore only where the oceanic drop-off is close to the coastline.

**Status in the Gulf.**—During the Fritts aerial surveys of the early 1980s, striped dolphins were regularly noted in the eastern Gulf of Mexico around southern Florida, but only occasionally in other areas of the Gulf. Striped dolphin distribution in all other areas of the world, however, indicates an affinity to waters deeper than 200 m (656 ft.), and it is likely that the sightings in more shallow water were misidentifications of Atlantic spotted dolphins (whose young are not spotted and have a prominent spinal blaze like striped dolphins).

The National Marine Fisheries Service estimates striped dolphin abundance at about 5000 animals in the northern waters of the Gulf. There appears to be a concentration of striped dolphins in the eastern part of the GulfCet study area, near and over the DeSoto Canyon region just east of the Mississippi Delta, and recent estimates there range as high as 2200 animals.

*Delphinus delphis* Linnaeus, 1758, Short-beaked Common Dolphin

**Names.**—The scientific name is derived from the Latin *delphinus,* "dolphin," and the Greek *delphis,* also "dolphin." Spanish: delfín comun de rostro corto.



**Description.**—Common dolphins have only recently been split into two species, a short-beaked form (*D. delphis*) and a long-beaked variety (*D. capensis*), and the two species are sometimes difficult to distinguish at sea. Both species have tall, slightly falcate dorsal fins, but the short-beaked common dolphin is somewhat more robust than the long-beaked species, which has a shorter (but still moderately long) beak and a more rounded melon.

Common dolphins average about 2.1 m (7 ft.) in length, with a maximum of about 2.6 m (8.5 ft.). Weight is generally about 75 kg (165 lb.), but may reach 136 kg (300 lb.). Males are typically a little larger than females.

Common dolphins are strikingly marked, with a dark charcoal back, white belly, and tan to ochre thoracic patch. This thoracic patch dips below the dorsal fin and combines with an area of streaked light gray on the tailstock to produce their most characteristic feature, the "hourglass pattern" on the side of the body. In the short-beaked common dolphin, the thoracic patch is relatively light, contrasting strongly with the dark cape. The flipper-to-anus stripe is weakly developed or absent. There are often light patches on the flippers and dorsal fin. The chin-to-flipper stripe does not approach the gape and narrows anterior to the eye. The lips are black, and a dark, distinct stripe runs from the apex of the melon to encircle the eye.

There are 41–54 pairs of small pointed teeth in each jaw (Table 2). At birth, common dolphins are 80- to 85-cm long.

**Behavior and Ecology.**—This dolphin is an oceanic species that is widely distributed in the tropical and warm temperate waters of the world. It occurs from nearshore waters to thousands of kilometers offshore. It is sympatric with the long-beaked common dolphin in southern California and off the west coast of Baja California, and perhaps in other parts of the world as well.

Large, boisterous groups of common dolphins often are seen whipping the ocean's surface into a froth as they move along at high speed. Herds range in size from about 10 to more than 10,000 animals. Associations with other marine mammal species are not uncommon. Active and energetic bow riders, common dolphins are very familiar to most seagoers in low latitudes. They are often aerially active and highly vocal. Sometimes their squeals can be heard above the surface as they ride bow waves.

Their prey consists largely of small schooling fishes and squid. In some areas, they feed mostly at night on creatures (associated with the deep scattering layer) that migrate toward the surface in the dark. Squid may form a more important part of the diet of short-beaked dolphins than of the long-beaked species.

**Distribution and Status.**—Common dolphins are distributed worldwide in tropical, subtropical, and warm temperate waters. In the western North Atlantic, they are known from Newfoundland, Iceland, Nova Scotia, and the coast of Massachusetts southward along the coast of North America to the Caribbean, and from South American waters at least to Venezuela.

Their appearance in the North Atlantic may coincide with the intrusion of warm waters in summer and fall; their numbers also increase off southern California in June, September–October, and January. These observations suggest that they may undergo seasonal movements induced by changing ocean temperature gradients (or by movements of their prey), but such migrational patterns have yet to be clearly documented.

These are offshore, pelagic dolphins that have suffered considerable losses in various parts of the world through direct and incidental catches. In the western North Atlantic, they were seen frequently off the northeastern coast of Florida prior to the 1960s; they have been conspicuously absent ever since, although they are still seen to the north and south of this area. Only short-beaked common dolphins have been recorded near Florida, with the closest documented record of the long-beaked species being from Venezuela.

**Status in the Gulf.**—Although both stranding and sighting records exist for the Gulf of Mexico, all museum skulls previously noted as *Delphinus* now have been reidentified as one or another of the *Stenella* species. Similarly, descriptions of *Delphinus* at sea may well be misidentifications, mainly of *Stenella clymene,* before that dolphin was rediscovered in 1981. No photographs are known to exist, and no alleged sightings have been made in the 1980s or 1990s in the northern Gulf. Observers from the Instituto Nacional de la Pesca from Mexico reported at least two winter 1995 sightings of common dolphins in the southern Gulf; however, this has not been confirmed. We conclude that although the genus *Delphinus* is not known definitively from the Gulf of Mexico, the animals will be described, perhaps both species, most likely in the southern waters that have not been surveyed adequately.



*Delphinus capensis* Gray, 1828, Long-beaked Common Dolphin

**Names.**—The scientific name of the genus means "dolphin" in Latin. The species refers to the type specimen locality at the South African Cape of Good Hope. Spanish: delfín comun de rostro largo.

**Description.**—Long-beaked common dolphins have the same basic body morphology as their short-beaked relative. Besides having longer beaks than their short-beaked relatives, they are slightly longer and more slender, and have a somewhat more flat appearance to the melon, which rises from the rostrum at a relatively low angle. Size is approximately that of *D. delphis.*

All common dolphins are characterized by an hourglass pattern on the side, forming a V below the dorsal fin. In the long-beaked species, the coloration appears somewhat muted compared to the short-beaked version. The thoracic patch is relatively dark, contrasting less with the cape. Generally, the flipper-to-anus stripe is moderately to strongly developed. The chin-to-flipper stripe fuses with the lip patch at or just anterior to the gape, and remains relatively wide ahead of the eye. The eye patch is lighter and less distinct than in the short-beaked species. Light patches on the dorsal fin and flippers are present only occasionally, and faint when evident.

The mouth is lined with 47 to 60 sharply pointed teeth in each tooth row (Table 2).

**Behavior and Ecology.**—Because of the recent discovery that common dolphins in the central Pacific represent two species (rather than only one, as was commonly thought), much of the biological information available for dolphins of the genus *Delphinus* cannot be applied reliably to one or the other species. The long-beaked form appears to inhabit more nearshore waters than the short-beaked species, generally occurring within 180 km (112 mi.) of the coast.

Herds of less than a dozen to several thousand animals are common. These dolphins are capable and willing bow riders, and often exhibit a great deal of aerial activity. One or the other type of common dolphin tends to predominate in the stranding record for southern California during any particular time period. In the year following the 1982–83 El Niño event, the long-beaked form was most common. A wide variety of schooling fishes and squids are taken as prey. In the northern Gulf of California, cooperative feeding techniques are sometimes used to herd fish schools.

**Distribution and Status.**—Because the two species of common dolphins have not been distinguished reliably until very recently, general distribution is assumed to be similar to that of *D. delphis,* but usually in less deep waters. They are present along the coast of Peru and the entire Gulf of California; they overlap with the short-beaked species along the west coast of Baja California, southern and central California, and probably other areas as well. In the Atlantic, they are known to occur off Venezuela, Brazil, and Africa, but detailed occurrence patterns are not known.

**Status in the Gulf.**—*Delphinus capensis* has not been recorded in the Gulf of Mexico.

*Lagenodelphis hosei* Fraser, 1956, Fraser's Dolphin

**Names.**—The generic name is derived from the similarity of this dolphin to species in the genera *Lagenorhynchus* (white-sided, white-beaked, Peale's, dusky, and hourglass dolphins) and *Delphinus* (common dolphins). The species name comes from Ernest and Charles Hose, who discovered the type specimen. Spanish: delfín de Fraser.

**Description.**—These are robust dolphins with very short beaks and small dorsal fins, flippers, and flukes. Adult males reach 2.7 m (9 ft.) in length and weigh at least 210 kg (460 lb.); females are somewhat smaller. At birth, these dolphins are 105–110 cm (3.4–3.6 ft.) in length.

These dolphins have a unique, chunky body shape with extremely short but well-defined beaks (adult beak length: 3–6 cm).

The dorsal fin is relatively small (maximum height: 22 cm) and slightly falcate to triangular. The flippers and flukes are typically dolphin-shaped, but smaller than in other species of similar size.

Although newborn animals appear to show a simple countershading with a dark gray back and white belly, older animals develop a complex color pattern. There is a dark cape, light gray or brownish sides, and a white belly sometimes with a pinkish hue. The appendages are all dark. A dark line similar to that in common dolphins (*Delphinus* sp.) and some species of the genus *Stenella* is found on the upper beak. There is a dark chin-to-flipper stripe, and a delphinid "bridle" (with a dark line from the melon apex surrounding the eye). A gray stripe of variable intensity and thickness runs from the facial area to the anus. In some animals, it is black and so thick that it merges with the flipper stripe and forms a "mask" along the side of the face. The side stripe is weakly developed in young animals, and reaches its height of expression in adult males. It also may be geographically variable, and appears to be less well developed in the Gulf of Mexico than, for example, in some parts of the Pacific.



There are 34–44 small, sharp teeth in each tooth row, and the skull has deep grooves along the palate (Table 2). Only common dolphins (*Delphinus* sp.) have palatal grooves as deep as Fraser's dolphins.

**Behavior and Ecology.**—These are very active animals, often seen traveling quickly, making splashy leaps, and turning the water's surface into a froth. Bow riding is rare in some areas but is a common behavior in the Gulf. They occur in large groups of several dozen to hundreds (or even thousands) of individuals. Fraser's dolphins often school with melon-headed whales. These animals feed primarily on small mid-water fish and squids.

The reproductive biology is not well known. Gestation is a little over a year and calves are born year-round, possibly with seasonal peaks in some areas. Males off Japan become sexually mature at about 7 to 10 years of age, and females at 5 to 8 years. The lactation period is not known.

**Distribution and Status.**—These are tropical and subtropical animals, occurring in deep water beyond the continental shelf edge. In some areas where deep water is found nearshore, they can be seen close to land. In many oceanic areas in the tropics, they appear in relatively low densities. They are common near some island groups, however, such as the Philippines. Long-range migrations are not thought to occur.

**Status in the Gulf.**—There are only a handful of records for Fraser's dolphins from anywhere in the Atlantic Ocean; however, more are from the Gulf of Mexico than anywhere else. The first record in the Gulf was a 1981 mass stranding in the Florida Keys. Until 1992 when the first sightings occurred, this was the only record for the Gulf. From 1992 through 1997, there were five sightings in offshore waters of the northern Gulf of Mexico (all associated with the GulfCet surveys) and at least three strandings in Florida and Texas. The sightings in the northwestern part of the Gulf were in waters centering around 1000 m (3,280 ft.) deep, with group sizes of 17, 22, 44, and 45 animals. Interestingly, Fraser's dolphins were associated with melon-headed whales in four of the five sightings. The estimate of numbers for the area is incomplete due to low sample sizes, but could be as high as about 1000 animals.



## ORDER SIRENIA, MANATEES AND THE DUGONG

The sirenians are large, plump, torpedo-shaped marine mammals adapted to live in bays and coastal waters of the world's tropical regions. Their closest living relatives are the elephants and hyraxes, and they resemble ungulates more than other types of marine mammals. There is only one species of sirenian in the Gulf, the West Indian manatee, which is now listed as a rare and endangered species.

### FAMILY TRICHECHIDAE, WEST INDIAN MANATEE

This family includes three species, one of which, the West Indian Manatee, occupies the coastal waters and some connecting rivers from Virginia around the Gulf of Mexico and the Caribbean Sea to eastern Brazil, the Orinoco basin, and the Greater and Lesser Antilles. This species inhabits shallow coastal waters—bays, estuaries, lagoons, and rivers. It utilizes both saltwater and freshwater, although there may be a preference for the latter.

*Trichechus manatus* Linnaeus, 1758, West Indian Manatee

**Names.**—The generic name is Greek for "to have hair" (*trichos,* "hair" and *ekhö,* "have"), probably in reference to the few hairs and sensory bristles on the face. The specific name is from the Carib *manati,* for "breast," referring to the axillary teats (and perhaps comparison to mythical mermaids). Other common names include Florida manatee, Caribbean manatee, and sea cow. Spanish: vaca marina del Caribe or manatí del Caribe.



**Description.**—The West Indian manatee is a rotund, massive animal with a horizontally flattened and rounded tail. Adults are 2.5–4.6 m (8–15 ft.) in length and weigh 200–600 kg (440–1320 lb.), with the highest recorded weight being 1650 kg (3630 lb.). The head is relatively small and squarish in profile with numerous bristles on the snout. The skin is up to 5.1-cm (2 in.) thick, heavily folded or wrinkled, and often heavily scarred from collisions with powerboats. The flippers are rounded and paddlelike and ear pinnae are absent. Coloration is medium gray or brown and the animals are often heavily covered with algal growths or encrusted with barnacles. Infants are much darker.

**Behavior and Ecology.**—Manatees are opportunistic, aquatic herbivores that feed exclusively on aquatic vegetation such as turtle grass (*Thalassia testudinum*), manatee grass (*Syringodium filiforme*), and water hyacinth (*Eichhornia crassipes*). Captive animals have been fed lawn grass (*Poa* sp.), dandelions (*Taraxacum officinale*), sow bread (*Sonchus oleraceus,* a thistle), palmetto fronds (*Sabal palmetto*), garden vegetables, and commercial fruits. Wild animals seem to prefer submergent vegetation, followed by floating and emergent species; food plants, however, appear to be selected indiscriminately within these classifications relative to availability. Manatees consume 29.5–50.0 kg (65–110 lb.) of food per day. They often feed extensively on floating rafts of water hyacinth and may pose some benefit in helping to keep commercial waterways from being clogged by this fast-growing plant.

Manatees may occur in loosely knit groups but are not gregarious by nature. Breeding and calving occur year-round, but is biased toward spring–summer; the gestation period is estimated at 12 to 13 months. Newborn manatees are about 1.2-m (3.5 ft.) long at birth and weigh 18.1–27.2 kg (40–60 lb.).

**Distribution and Status.**—West Indian manatees are found in rivers, estuaries, and coastal areas of the tropical and subtropical regions of the New World Atlantic, ranging from the southeastern

United States southward along the Central American coast to the West Indies and northern coast of South America to central Brazil. Their range in the United States is confined largely to peninsular Florida and the coast of Georgia, but a few records are known from the northern Gulf of Mexico around to the mouth of the Rio Grande.



Manatees inhabit both shallow freshwaters and saline waters, and occasionally may wander into the open ocean, although rarely more than several miles from shore. During winter months, they may move into warm freshwater habitats, and often congregate around warm freshwater springs and warm water effluents discharged from industrial plants. They have a slow metabolism and high rate of heat loss and simply cannot tolerate temperatures much below 18°C (64°F). In summer, they disperse back into warm saline waters and reinhabit the northern areas of their range. Such movements probably account for the occasional sightings of manatees in the northern gulf west of Florida.

In Florida, *T. manatus* numbered at least 2600 individuals during aerial surveys made of the shoreline and power plant areas, when many animals were aggregated there during a period of cold weather. This is a "minimum population" count, and cannot be extrapolated to an estimate of overall numbers. The USFWS has classified the West Indian manatee as endangered.

**Status in the Gulf.**—Excluding the Florida coast, manatees are rare in the Gulf of Mexico. Only fifteen records have been reported from the northern Gulf prior to 1970. Since 1970, there have been other reports from this region, including sightings off Texas, Louisiana, or Mississippi almost every summer.

Apparently, manatees were often observed in the Laguna Madre of south Texas around the first decade of this century. Subsequent reports, however, were rare. One individual washed ashore dead near Bayside, Texas, in 1928, and another dead animal was found near Sabine Lake (Texas–Louisiana border) in 1937. *T. manatus* was never common along the Texas coast, and these reports may have been from strays in Mexican waters farther to the south.

Records from Louisiana and Mississippi, however, have increased in the past 15 years. This may indicate a gain in population numbers along the Florida coast, as manatees are thought to migrate northward in the Gulf during the warmer summer months. Mana-

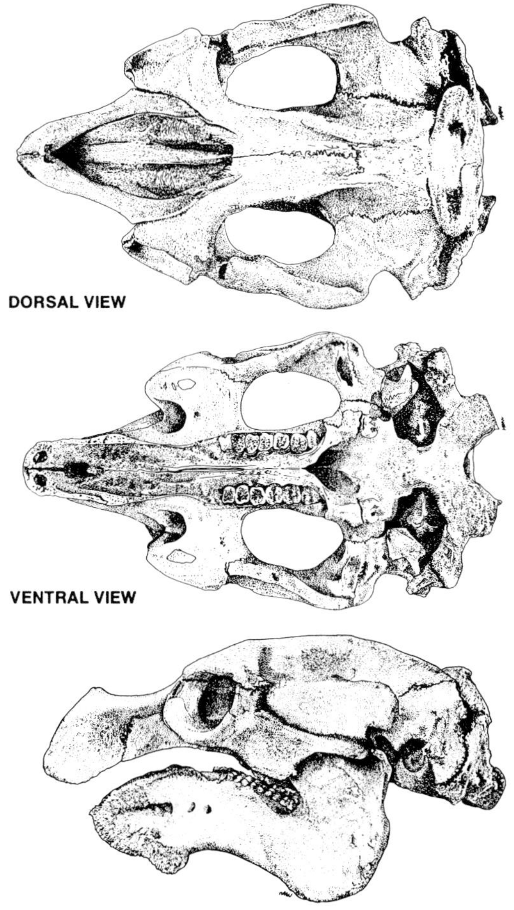


*38. West Indian manatee,* Trichechus manatus

tee numbers apparently increased in Florida during the 1970s and 1980s, but may be stable or declining at this time. One thing is clear—recent large increases in habitat destruction and power-boating do not bode well for the long-term health of populations in U.S. waters. In winter of 1997, a massive die-off of hundreds of manatees in Florida waters was linked to an outbreak of morbillivirus, a distemper-like virus. It is not clear whether this viral infection gained hold due to polluted waters or some other anthropogenic effect.

**Remarks.**—Sluggish and easily captured, West Indian manatees once were exploited extensively as a food source. Although they are now protected, they still face extensive losses, especially from collisions with high-powered speedboats. Additionally, habitat loss to land development and channelization continues to pose problems. Ironically, construction of power plants and industrial plants apparently has been beneficial, creating new warm water habitats that may be preferred by manatees during winter. The downside to these artificial habitats is that they keep considerable numbers of manatees north of their historic wintering grounds, and could subject them to unusually cold conditions if a power plant should temporarily shut down.



## ORDER CARNIVORA

### *Suborder Pinnipedia, Seals, Sea Lions, and Walrus*

This suborder of aquatic mammals occurs along ice fronts and coastlines, mainly in polar and temperate parts of the oceans and adjoining seas of the world, but also in some tropical areas and certain inland bodies of water. Only one species occurred naturally in the Gulf of Mexico, and it became extinct earlier this century. Another species has been introduced into the region and may occur occasionally in the feral state. Pinnipeds are less modified for aquatic life than the wholly aquatic cetaceans.

The two families formerly represented in the Gulf include the sea lions (Otariidae) and the true seals (Phocidae). In the otariids, the hind flippers can be turned forward to help support the body; thus all four limbs can be used to travel on land. Forward motion in water is effected mainly by the powerful front flippers that scull or paddle in a fashion similar to that of sea turtles. In the phocids, the hind flippers cannot be moved ahead, and the animals must wriggle and hunch, much like giant caterpillars, to travel on land. Forward motion in the water is gained by moving the posterior one-third of the body and the hind limbs from side to side, in contrast to the up-and-down motion of cetaceans.

We list the West Indian monk seal in honor of their former status on the fringes of the eastern Gulf of Mexico. The California sea lion (*Zalophus californianus*), however, was introduced to the Gulf by humans and no verified sightings have been reported since 1972.

#### FAMILY PHOCIDAE, TRUE SEALS

*Monachus tropicalis* (Gray, 1850), West Indian Monk Seal

**Names.**—The generic name is from the Greek *monakhos,* referring to the monklike smooth, round head. The species name refers to the tropical nature of its range. Other common names include Caribbean seal and West Indian seal. Spanish: foca monja del Caribe.

**Description.**—The West Indian monk seal, now extinct, belongs to the pinniped family Phocidae. These seals were characterized by the absence of ear pinnae, hind limbs that were turned posteriorly and modified into flippers, and small front flippers that were helpful in steering underwater.



Adult West Indian monk seals averaged about 2.3 m (7.5 ft.) in length but Townsend (1906) reported one old female, captured about 8.1 km (5 mi.) from Key West, Florida, that measured 2.7 m (9 ft.) in length. Pelage coloration was brown tinged with gray dorsally, lighter on the sides, and pale yellow or yellowish white ventrally.

**Behavior and Ecology.**—Virtually nothing was learned about the life history of this species before its extinction. Apparently, the young were born in early December, based on several females killed in the Triangle Keys during this time carrying well-developed fetuses. The animals were remarkably sluggish and allowed people to come among them without great alarm; thus many could easily be killed. No doubt, this lack of suspicion and fear contributed to their extinction. Their diet probably included fish and mollusks.

**Distribution and Status.**—West Indian monk seals were the only seal native to the Gulf of Mexico. They were distributed tropically and limited to the Gulf of Mexico coast, the Yucatan peninsula, western Caribbean Sea, the Greater and Lesser Antilles, the Bahamas, and the Florida Keys. Their remains have been excavated from archaeological sites in coastal Texas, supporting their probable occurrence in the far western Gulf.

The Alacranes Islands and Triangle Keys off the coast of Yucatan were apparently their last remaining stronghold. Residents of Carmen, Yucatan, reported seals in the Alacranes as late as 1948. The last authenticated record was an observation in 1952 of a small colony on Serenilla Bank, which is in the Caribbean Sea midway between Honduras and Jamaica.

Aerial surveys were conducted in 1973 around the islands and atolls off Campeche, Yucatan, Quintana Roo (Mexico), Belize, Honduras, Nicaragua, and the central Caribbean to Jamaica. At every island group visited, either fishing vessels (or shrimp trawlers) at anchor or fishing crews and their shacks (or the remains of abandoned camps) on shore were observed, with no indication of the West Indian monk seals. It was concluded that the monk seal had been extinct since the early 1950s. Numerous expeditions since that time have come to the same conclusion.

The early decimation of the West Indian monk seal was brought about by overhunting because they were the best source of oil in the southern islands during colonial times. Seals were slaughtered so persistently that they were rare by 1851. The subsequent extinction of these seals can be attributed directly to human disturbance, especially fishing. The most remote habitat of monk seals had been invaded by fishing crews that were prone to kill seals as competitors. Because monk seals evolved in island environments in which they had no natural enemies ashore, they were inherently tame and thus easy victims.



**Status in the Gulf.**—The last specimen of *M. tropicalis* from the Gulf was taken in 1922 near Key West. After that, sightings were made along the Texas coast, one in 1932 and another possible one in 1957 near Galveston, which was reported in a local weekly newspaper. The latter record, however, cannot be verified and probably represented the sighting of an escaped California sea lion. Archaeological sites on the Texas Coast have provided some additional evidence for the seals' former occurrence off Texas, but it is possible that these remains also could have been trade items obtained by natives elsewhere.

# Chapter 6

# Worldwide Status and Conservation

A SPECIES-BY-SPECIES ACCOUNT of the present population status and numbers of marine mammals in different parts of the world is not possible in the space available. Even if this could be accomplished, it would be a very incomplete exercise because of the difficulty in obtaining accurate population-size estimates for all but the coastal species. Several broad statements, however, can be made about the conservation status for the major groupings of marine mammals.

## ORDER SIRENIA

Wherever they occur, sirenians are not thriving as species or populations. These are coastal animals and therefore, inevitably come into contact with humans over much of their ranges. In Florida, the West Indian manatee was reduced by hunting in previous centuries, and today is threatened by habitat destruction, periodic viral infections, and being hit (often killed) by the hulls, drive units, and propellers of swiftly moving powerboats.

The small Amazon manatee of the giant river systems (the Amazon and Orinoco) of South America is threatened with capture for subsistence use in almost all parts of its range, and by habitat degradation near human population centers. Two of us (BW and TAJ) spent time in the upper Amazon along the Marañon and its tributaries in Peru, and wherever they traveled, even into the smallest byways accessible only by dugout canoe, there was evidence of traps and nets set for manatees. When a manatee was encountered, it proved to be extremely shy of human presence; with a furtive, shallow blow at the surface, it would disappear.

Such behavior is a sure sign of hunting by humans. Where hunting does not take place, such as in several protected lakes in Brazil, Amazon manatees are curious and playful around humans, just like the Caribbean manatees in Crystal River, Florida. The African manatee is eaten, and its numbers and state of well-being are essentially unknown at this time.



In northern Australia and parts of Asia, dugongs are killed for their flesh and fatty oils; the small tusks of males are ground into a fine powder for medicinal purposes. These tusks are reputed to be an aphrodisiac and enhance sexual potency in human males, a dubious claim for body parts that has caused the demise of many a mammal. Dugongs appear to be declining throughout their range from the Red Sea to their last stronghold in Australia; accurate information, however, is not available. Hopefully, they will not suffer a fate similar to their extinct giant cousin, the Steller's sea cow, which was wiped out by hunters within 28 years of its discovery in 1741 by "Western science."

## Other Marine Mammals, Polar Bears and Sea Otters

Polar bears are thriving in most parts of their range in northern Alaska, Canada, Greenland, and Siberia. They apparently have a sufficient supply of seals and fishes upon which to feed and even take advantage of the local dump in Churchill, Manitoba, Canada. Their savior for the time being is the northern environment, where human influence is still relatively small. Eskimos, however, do kill polar bears for fur and meat throughout most of their range, and a recent sport of polar bear hunting has taken a strong hold for southern adventurers. For a fee (generally U.S.$10,000–15,000), the hunter is "guaranteed" a polar bear as a trophy. Such a "sport" is always worrisome for a large, upper trophic level carnivore that reproduces slowly. If thousands of such hunts take place in the future, as they do now from helicopters and rapidly moving skimobiles, then even the polar bear populations of the Arctic may no longer be safe, and the species could become threatened. The American bison and all of the large carnivores of Africa are other sad examples of sport overhunting.

Sea otters were severely depleted for their pelts in the nineteenth and early parts of the twentieth century all along their range, from Baja California, Mexico, up through western Canada and the Alaskan archipelago, and to northern Japan. They were thought to be extirpated south of Alaska, but have made a remarkable recovery; there are an estimated 1200–1400 sea otters off central California at this time. There are indications, however, that the population is decreasing. Quite a few sea otters are killed in coastal gill nets, which may be a major factor in their apparent decline. Sea otters also are taken by killer whales and sharks. Some species of sharks, especially the great white, have increased off California and farther north, probably due to a recent increase in prey such as elephant seals, sea lions, harbor seals, and other small marine mammals. Killer whales off Alaska may be relying on sea otters to a greater extent than previously because of a recent decline in Steller's sea lions and northern fur seals. It also is possible that bottom food taken by sea otters is becoming scarce in some areas. This is a complicated issue, and hopefully, sea otters are not in danger of extinction, either as populations or the species as a whole. Little is known about the status of the marine otter, the chungungo off South America, although it has been depleted and totally extirpated from some areas.



## ORDER CARNIVORA

### *Suborder Pinnipedia*

Some species of pinnipeds are flourishing while others are imperiled. Most sea lions and fur seals were hunted extensively for their pelts, and many true seals were taken for oil during the past several centuries. Almost all have made remarkable recoveries. Elephant seals, like sea otters, were almost completely wiped out in the North Pacific, and badly depleted in the southern hemisphere as well. They now have returned to pre-exploitation numbers despite having gone through a "population bottleneck" (of perhaps as few as only several dozen individuals) that left them extremely low in genetic variability. Nevertheless, they are thriving, causing those who breed endangered species for potential reintroduction to the wild to question the need for high genetic heterogeneity in all mammals. It is not known, of course, if this low variability in gene expression could make the elephant seals eventual victims of climate or other unforeseen changes that they might have "weathered" with greater genetic variation.

Neither Weddell nor crabeater seals, which feed largely on euphausiid crustaceans (krill), were ever hunted as extensively as elephant seals. They now have reached very large numbers, apparently because the anthropogenic decline of blue and fin whales in antarctic waters has left huge resources of krill available to them. Indeed, crabeater seals are now the most numerous of marine mammals, with well over 20 million estimated in the antarctic. Since the seals reproduce much faster than the great whales, a shift in ecosystem balance has taken place; the seals are now the dominant marine mammal krill feeders in the antarctic, which most likely will prevent whales from recovering to pre-exploitation levels in the area.



Northern fur and harp seals in the northern hemisphere have been killed yearly in well-managed hunts designed to keep the population level stable, or increase slowly. The harp seal hunt is now strictly regulated, not so much because of overhunting, but more because of public outcry over killing young and defenseless (i.e., cute) pups. A traditional Alaskan native fur seal hunt on the Pribilof Islands has been terminated because fur seals have declined drastically in numbers. This was not due to the hunt but probably because of accidental entanglements in salmon and other fishing nets used on the high seas of the North Pacific Ocean. Human overfishing of the very fishes and squid that are prey of the fur seals is probably another major part of the problem. Steller's sea lions have also dropped in numbers, most likely for the same reasons. Huge drift nets had plied the North Pacific Ocean from the 1950s until recently, and precipitous declines in salmon, squid, and many other stocks of near-surface creatures have resulted.

Monk seals, the only true tropical group of seals, are in desperate trouble. They have died out in the Caribbean (the last confirmed reports are from the early 1950s), and are faring very poorly in the Mediterranean Sea and the Hawaiian archipelago. Past hunting for food and present habitat destruction of pupping beaches are the major culprits leading to their demise. There is now an unexplained paucity of Hawaiian monk seal females; males have begun to mate disruptively with all available adults and youngsters, often mobbing a pupping beach en masse and killing newborns in the process. The U.S. government is experimenting with translocations of males and attempting to set up foundling populations in decimated areas on the outer Hawaiian atolls. To our knowledge the West Indian monk seal is the only species of marine mammal to become extinct in this century, but the Mediterranean and Hawaiian monk seals soon may follow if quick and decisive protection of the animals and their habitats is not instituted. In 1997, a massive die-off due to a viral infection further reduced the Mediterranean population, and put them even closer to the sad brink of extinction.

Walrus, the third group of pinnipeds, appear to be surviving reasonably well in the arctic. Eskimos have traditionally hunted them, utilizing all parts of the body—the hide for tents, meat for food, bones for construction material and utensils, tusks for adzes, intestines for rope, stomachs as sealing floats, and even the penis bone (*os penis* or baculum) as a hunting club. The ivorylike tusks and bacula are also used as carving and scrimshaw (the art of etching drawings into hard bone and then blackening the etched lines) material, and there is now a thriving business of selling these often beautiful items of native art to tourists. Unfortunately, this has led to localized depletions of walrus in some areas, with Eskimo hunters killing entire beached groups of animals (both males and females have tusks) with so-called large-gauge "elephant" (or "polar bear") rifles and cutting out only the tusks and os penises with chainsaws. This newly developed habit is a far cry from the ancient one of utilizing the entire animal. One of us (BW) once encountered more than one hundred bloated carcasses that were slaughtered (as described above) and floating off King Island in the northern Bering Sea—a stomach-churning experience that would preclude his ever buying or trading walrus ivory again. As unfortunate as such events are, however, the overall status of walrus appears to be good, probably because they spend so much of their time on remote ice floes away from human habitation.

## ORDER CETACEA

Large whales were hunted extensively in the past several centuries. At first, it was the right whales that were almost extirpated. They were relatively slow, floated when killed, and yielded enormous quantities of oil. They were indeed the "right whales" to hunt. Then bowheads, sperm whales, and gray whales were taken. Until the sixteenth century, gray whales, the so-called "scrags" in the early European whaling literature, were also taken in the North Atlantic. Sadly, two entire populations were extirpated, on both the coasts of the Americas and Europe, and today no nearshore gray whales survive in the Atlantic.



In the twentieth century, huge factory ships made the efficient taking of thousands of whales per season possible, and humpbacks, blues, fins, seis, and finally the smaller minke whales, were taken in large numbers. Several stocks were wiped out or reduced to nearly zero (North Atlantic and Korean gray whales, right whales in all oceans, and eastern Atlantic bowheads), but no species of great whale has become extinct as a result of these activities in modern times. This fortunate fact was not due to human foresight, but was a result of the difference between economic extinction and biological extinction. When it became economically unfeasible to search for and find particular whales, the industry shifted to other species.

A temporary whaling moratorium, overseen by the International Whaling Commission based in Cambridge, England, prohibits all but a few regional and relatively small whaling operations. This moratorium came into effect in 1988 and was renewed in 1992. Hopefully, it will be renewed again in the future; blue, humpback, right, and bowhead whales, as well as several other species have been greatly reduced over most parts of their former ranges. About 50 bowhead whales are still taken by Alaskan Eskimos each year, and 160 gray whales (which have made a dramatic recovery after near extinction 70 years ago) are taken by Russian Eskimos in far eastern Asia. Small "mom and pop" operations still take whales off the Philippines, the Azores, Sri Lanka, certain islands of Indonesia, and occasionally, small island nations such as those in the Caribbean. Pirate whaling also continues in unknown, but probably not extremely high, numbers. Today, Norway, Iceland, and Japan continue to take limited numbers of minke whales. Although there has been a public outcry over this throughout most of Europe and North America, this whaling does not harm the relatively large populations of minke whales. It thus becomes an "animal rights" and not a conservation issue. Nevertheless, those opposed to even low levels of minke whaling point out that such activities keep the industry alive (i.e., whaling boats and gear in commission, whalers employed), and therefore could lead to greater numbers and species being taken again in the future. Indeed, Norway has vowed to increase minke whaling and also hunt the much larger fin whales. While the Norwegian government claims that all whale meat is used for local whale-town purposes (so-called "subsistence whaling"), illegal whale-meat shipments, often labeled "shrimp," have been seized in ships and airports (bound for Japan); several Norwegian companies admit to stockpiling huge quantities of frozen blubber for export to Japan should regulations in Europe change. Recent genetic studies of "minke whale" meat sold in Japanese markets has shown it to be that of humpback and other whales, demonstrating that species banned from hunting are also being taken.

The issue of whaling is complicated. For example, in our culture we tend to see nothing wrong with Alaskan Eskimos killing a limited number of bowheads each year. One of us (BW) spent fifteen summers with the Eskimos and understands the importance of the hunt. It is not just for obtaining food, but also a source of valuable tribal pride; the hunt represents old customs in a society rooted in two very different cultures often at odds with each other. Yet, the United States arguably has been the driving force (through representation in the International Whaling Commission and international economic sanctions and threats of sanctions) in reducing large-whale hunting to nearly zero. Representatives of Japan, Iceland, and Norway argue that their whaling history goes back as far as indigenous hunting by other cultures, and that they should be afforded the same respect and rights as the Eskimos. The validity of such arguments, and the social (at times, economic) issues involved are not easily understood, but suffice it to say that the large baleen whales and the sperm whale appear to have a brighter future today than they did twenty years ago.

The most serious anthropogenic threats are to some of the smaller toothed whales that, until very recently, have not been regulated by any concerted national or international efforts. Danger to these animals comes from three major human sources: intentional hunting, accidental demise due to fishing operations, and habitat destruction (including pollution).

Intentional hunting of toothed whales is practiced in many areas. The Japanese fishing industry kills thousands of Dall's porpoises and striped dolphins every year for meat, and periodically kills pilot whales, bottlenose dolphins, and other toothed whales that compete with their fishing activities. Indeed, dolphins are well known to interfere with several of Japan's fisheries by scaring fish away from nets or lines and directly taking fish from gear. Many fishing crews head for port when pods of dolphins arrive, and some shift to demersal or other often less valuable fishes or invertebrates.

For centuries, the Japanese have practiced a type of small cetacean catching called a "drive fishery"—boats form a semicircle around a school of dolphins and each boat lowers at least one metal pipe flared at the bottom (like a trombone) that is repeatedly banged. The cumulative noise of the pipes from all vessels (as many as 800 boats drive a school of several thousand dolphins) effectively drives the dolphins to land (usually a bay) where they are slaughtered in shallow water. Because such an efficient technique takes the entire genetic stock of a huge school, such dolphin "whaling" has had a devastating effect on many of Japan's nearshore cetaceans. Villages that 20 years ago had yearly drives numbering thousands of animals, have only on average one drive every several years, with sometimes as few as 30 or 40 dolphins captured.

The exact number of cetaceans remaining in Japanese waters is not known, but some amount of conflict between human fisheries and the existence of dolphins still exists around several islands. Incidentally, because an entire school can be caught by the drive fishery method and since a cadre of excellent Japanese scientists has taken advantage of this to gather data from the dead animals, some of the most complete life history information of any cetacean species is available for several dolphins and pilot whales from Japanese waters. Of course, this is a consequence and certainly not a justification for the wholesale slaughter.

Peruvian fishermen until very recently took thousands of dusky dolphins, Burmeister's porpoises, and other toothed whales yearly for crab-pot bait meat and human consumption. Most of this activity appears to have been curtailed by government action, and some former dolphin hunting boats are taking tourists out to see the live animals in nature. Fishing people in Sri Lanka and other areas in the Indo-Pacific kill dolphins for much needed protein in that impoverished part of the world. Narwhals are hunted extensively by Canadian Eskimos for the males' tusks, prized by many eastern cultures for purportedly enhancing virility, and their putative aphrodisiac and medicinal powers. Dall's porpoises especially suffer from intentional hunting, and populations in the northwestern Pacific waters have become drastically reduced.

Accidental kills are every bit as, if not more, devastating as the intentional ones. For over 30 years, intensive purse-seine tuna fishing in the eastern tropical Pacific has depleted spinner, spotted, and common dolphins (which often have tuna associated with them). Fishing crews set large nets around dolphins in hope of catching the tuna swimming below, and dolphins often became entangled in the nets and suffocated to the unfortunate number of well over 100,000 a year in the 1970s and 1980s. This intensive and long-term pressure has drastically reduced the formerly huge herds of small oceanic dolphins in the Pacific, not unlike grazing ungulates such as bison and wildebeest on land.

A U.S.- and European-supported boycott of tuna meat caught by purse seining in areas in which tuna and dolphins associate has strongly reduced the slaughter, but several thousand dolphins a year still die in nets set by some Central and South American countries. The issue is a complicated one because the demand for yellowfin tuna, a species associating with dolphins in the eastern tropical Pacific, is high in Europe and North America. Tuna caught by purse seining around dolphins has been illegally mislabeled by several non-U.S. nations and sold as "dolphin-safe" in Europe; however, it is not known how widespread this black market activity is. Because of economic pressure, there have been repeated attempts in the U.S. Congress to reverse the "dolphin-safe" aspect of the American tuna industry by once again allowing U.S. tuna nets to be set around schools of dolphins; those attempts have some validity because well-practiced netting and dolphin-separation tactics are known to be highly effective in reducing dolphin kills. The argument is, "let us do it right, not somebody else with perhaps less control." While there may be some truth to this, it is also true that foreign vessels have recently reduced their own kills to a substantial degree. The problem remains unresolved, and although political regulations will likely be changed in the near future, we believe that at least several thousand dolphins a year will be killed by tuna-dolphin purse seining for a long time to come—hopefully, however, at numbers that are sustainable in biological population and species terms.



Dall's porpoises of the North Pacific, harbor porpoises of Canada, Europe, and the United States, Burmeister's porpoises and dusky dolphins off Chile and Peru, the Gulf of California harbor porpoise of northern Mexico, Hector's dolphins of New Zealand, and several other mainly nearshore species are dying in untold numbers in gill nets set for salmon, croakers, squid, and other commercially valuable food sources. Dusky dolphins present a particularly sad story. At first they were caught accidental to netting activities near shore, and their meat was used to bait crab pots. Then, the animals were actually targeted for hunting. Then it was discovered that dusky dolphin meat was fit for human consumption and the directed fishery increased.

The situation with respect to Hector's dolphins presents a relative "success story." While it was being heavily decimated in the 1980s by nets in its nearshore habitats off South Island, New Zealand, effective governmental regulation of when and where nets could be set, drastically reduced the kill of this New Zealand endemic and appears to have assured its survival, at least in the foreseeable future.

The Gulf of California harbor porpoise, the vaquita (*Phocoena sinus*), has been hard hit, especially by nets set for a large (itself endangered) sea bass locally known as the totoaba. All indications are that the vaquita and the sea bass are becoming rapidly extinct in their restricted range (the northern Gulf of California near the mouth of the Colorado River).

All river dolphins (families Platanistidae, Iniidae, and Pontoporidae) are also caught in nets and other fishing gear. The almost blind Indus and Ganges River dolphins, which are considered two separate species, are now found only in very low densities in the waterways of the heavily populated Indian subcontinent. The Chinese river dolphin, or baiji, with perhaps only 50 to 100 individuals left, is in imminent danger of extinction; the baiji and vaquita vie for the unenviable position of most endangered cetacean.

This leads to a discussion of general ecological problems. The baiji and vaquita did not reach their low population status solely due to being killed incidental to fishing operations. They both live in environments that have been drastically altered by humans. The baiji's habitat, the Yangtze River, is the major highway into central China. It is heavily polluted from industry and agricultural chemical runoff. Its tributaries and byways have been altered by dams, dredging, weirs, and the river traffic of barges and ferries; yet its water and productivity help support over 15 percent of the world's population. The vaquita lives in an environment heavily silted by damming of the Colorado River, and is heavily exploited by an intensive shrimp fishing industry. Similar, but not quite as drastic, habitat alterations exist for the Indus, Ganges, and Amazon River dolphins, the harbor porpoise in many parts of its range, the bottlenose, Irrawaddy, and hump-backed dolphins, and others.



Especially detrimental, and insidious because of difficulties of measurement, are numerous unseen pollutants such as heavy metals, organochlorines, and other toxic concoctions produced by humans. Certain levels of DDE, a breakdown product of DDT (still used by many developing countries to control agricultural pests, and one causing eggshell thinning in birds, spontaneous abortions in high body loads, and immune dysfunctions), PCBs (compounds used as transformer coolants worldwide), and other chemicals lower reproductive capabilities, affect immune responses, and therefore, could threaten entire populations of marine mammals.

Dolphins, porpoises, and pinnipeds are "higher level" meat eaters, which means they tend to feed on prey that has already passed through several trophic levels—from zooplankton feeding on phytoplankton, invertebrates and small fishes feeding on the zooplankton, larger fishes and squid feeding on those, and so on. At each level, there is a bioaccumulation of substances that the bodies cannot readily digest or excrete, and the bioaccumulated effects can and do result in toxin loads in mammals many times the original concentration in the sea. Because many toxins are transported through the placenta and the mother's milk to fetuses and newborns, respectively, young marine mammals start life with a disadvantageously high body toxin load.

Several dramatic examples of the effects of pollution illustrate this point. In the late 1980s, about 20,000 harbor and gray seals died in northern European waters. The virus responsible was eventually identified as phocine distemper virus, or PDV, a class of so-called morbilliviruses. The story appeared to be over with many people suspecting, although without proof, that the thick, toxic brew of industrial, chemical, and agricultural pesticides and herbicides in the North Sea and other nearshore European waters contributed to this massive epidemic. Recently, research in Holland directly and convincingly linked impaired immunity with bioaccumulated environmental contaminants in seals; findings indicated that pollution probably contributed to the severity and extent of the outbreak, and is now adversely affecting the immunocompetence of marine mammals in many areas of the industrialized world.

Just as the northern epidemic of seals was abating, striped dolphins of southern Spain began to wash up on shores—dead. This trend appeared next on the Mediterranean coasts of Spain and France, and continued to Italy and Greece. Obviously something was killing these dolphins by the hundreds and into the low thousands, with the disease spreading west to east in the enclosed Mediterranean Sea. Finally, a strain of distemper virus (morbillivirus) related to that of the seals was isolated, and one investigator found that PCB levels tended to be twice as high in affected than in "healthy" (accidentally caught by net) members of the population. The southern outbreak has run its course, but it (or a variant) could reappear without warning. In spring of 1997, Mediterranean monk seals also died en masse, with the major culprit being a new strain of morbillivirus that possibly evolved from the seal distemper virus. In all these cases, the suspicion is present but without direct proof, that biotoxins in the environment are increasing susceptibility to viral outbreaks. From 1987 to 1988, at least 740 bottlenose dolphins died off the eastern United States coast. While no single agent of mortality was identified, possible infection due to morbillivirus and decreased immune responses cannot be ruled out.

The situation in the Black Sea, which adjoins the Mediterranean, is especially tragic. In the 1940s and 1950s, Russian scientists reported common dolphins in schools numbering tens of thousands. Many bottlenose dolphins came close to shore, and there were so many harbor porpoises that a lucrative Turkish fishery for harbor porpoise meat flourished. Today, the Black Sea is almost dead. Heavily overfished and extremely polluted from the Danube

River, which runs through some of the most industrialized nations on earth before flowing into this almost completely enclosed inland body of saltwater, the Black Sea is an extreme example of how poorly preserved some of our treasured world resources are.

There are many other examples of poisons in marine mammals. Extremely heavy toxin loads have been documented in white whales of the St. Lawrence Seaway, pilot whales north of England, all dolphins, porpoises, small whales, and pinnipeds caught off Japan and Taiwan, Indo-Pacific hump-backed dolphins and finless porpoises off Hong Kong, and even dusky and Hector's dolphins off the generally nonindustrialized islands of New Zealand. In the latter cases, the culprits involved various classes of agricultural pesticides. Unfortunately, even the high arctic and antarctic are affected by rising levels of global pollution, and recent evidence for Greenland harp seals and antarctic fur seals shows still low, but uniformly detectable, levels of polychlorine substances. The levels in the antarctic are still lower than those in the arctic at this time. Even the deep-feeding sperm whales of the North Atlantic have been found with abnormally high concentrations of heavy metals in their bodies. These large-toothed whales feed on fishes and large squid that are also high on the food chain, but it is not known if recent die-offs of sperm whales are related to such loads.



## The Situation in the Gulf of Mexico

Although the environmental situation in the Gulf of Mexico is not good, it is better overall than in the European North Sea, and probably a bit better than in the Mediterranean Sea. That is not to say that pollution does not flow as efficiently into the Gulf of Mexico as in some of the most polluted parts of earth. Indeed, there are high levels of water draining from the Mississippi and adjoining basins, and agricultural pesticide runoff is particularly heavy. Also, some of the most active petroleum hydrocarbon refining and production works on earth occur off Gulf of Mexico waters. In the north is the Houston–Texas City area on the shores of Galveston Bay. About 30 percent of all petrochemical refining and fully 50 percent of hydrocarbon product production (e.g., paints and plastics) in the United States occur around the shores of this bay. In the south are the industrial complexes of Tampico, Alvarado, and other Mexican cities.

There is good news, however, in all of this. First, the Gulf of Mexico is flushed better naturally than the shallow North Sea and the somewhat more enclosed Mediterranean. Second, strict (and strictly enforced) pollution regulations in the United States have measurably decreased toxin output from the Mississippi River, Galveston Bay, and elsewhere. The northern Gulf is now a cleaner place than it was 20 years ago, and this partial reversal of environmental degradation gives hope that all is not doomed to spiral inexorably downward. There is a need for more and better enforced regulations in some business sectors, and agricultural pesticides (which, after all, are specifically made to kill biological systems) especially need to be controlled more efficiently. Unfortunately, the situation in Mexico is less well controlled. Mexico has some of the best antipollution laws on earth, but enforcement is lax to nonexistent in many areas.

How does pollution that exists in Gulf of Mexico waters translate to the health of ecosystems and the health of marine mammals in this region? The two of course are intertwined. Problems seem to be occurring mainly in shallow waters near shore, and are thereby especially deleterious to the nearshore environment of the bottlenose dolphin. Recently, major die-off epidemics of bottlenose dolphins occurred along the Texas shore, with 201 animals stranding in 1990, 245 in 1992, and 286 in 1994. These die-offs compare to averages of 128 stranded dolphins since 1987 during nonpeak years. The increasing numbers can be interpreted in one of two ways; either dolphins are increasing in numbers and more die and come ashore naturally, or there is a recurring, and possibly increasing, problem that is producing higher mortalities. The first possibility is ruled out by Texas A&M University population studies made since 1990; there is no evidence of increasing numbers. The second explanation appears more likely. The 1994 die-off has implicated, at least in part, a strain of morbillivirus found in a number of animals. Almost all of the dolphins that stranded exhibited abnormally high amounts of nontoxic lesions, growths, and other abnormalities in parts of their bodies. Dr. Dan Cowan, a well-known pathologist at the University of Texas Medical Branch in Galveston, documented a litany of previously unknown abnormalities in these animals. Again, it is not known if high levels of artificial toxins are responsible for some of these deaths, but there is reason to be suspicious since high levels of heavy metals and polychlorinated products have been found in body tissues.



The dolphins and toothed whales of deep water simply have not been studied enough to form reasonable conclusions on the health of individuals, populations, or ecosystems. It is suspected, however, that some problems of chronic pollution exist there because global ocean toxins are rising in measurable fashion, even far from their point sources of input to the marine environment. Clearly, detailed studies on deepwater marine mammals in the Gulf need to be continued and relationships to environmental pollution explored at all available opportunities.

One obvious potential source of negative interaction between humans and marine mammals is that of offshore oil and gas exploration, development, and production activities. This has gone on in the Gulf for over 50 years, and effects on the toothed whales of these waters have been largely unstudied. In the arctic, one of us (BW) found measurable and repeated short-term effects of the oil industry on bowhead and white whales. Especially disruptive was the loud and abrupt "seismic pinging" that oil companies use to map the stratigraphy of the ocean floor in the hope of predicting profitable areas to drill. Seismic pinging has been implicated in chasing bowhead and white whales out of an area; recently, it was suggested that sperm whales also are harassed by these loud sounds that can travel to about 100 km (62 mi.) from their source vessel. Dolphins and whales, however, are also known to adapt and tolerate noxious noises that do not lead to danger, and it seems likely that these animals can survive with moderate degrees of oil industry activities on the high seas. Nevertheless, it is incumbent upon us to do the requisite monitoring of cetaceans near seismic vessels, oil rigs, and other industry related activities. This type of "behavioral monitoring" has occurred in oil productive areas elsewhere in U.S. waters, the arctic, and off the shores of California, but it has been curiously absent from the political agenda of needs for the Gulf of Mexico.

Another point of concern is that of a disastrous oil spill, like the huge 1989 *Exxon Valdez* spill in southern Alaska. Oil spills, which occur sporadically and in unpredictable areas of earth, are known to be drastically detrimental to those surface-breathing animals that use feathers or fur to thermoregulate, including marine birds, many of the pinnipeds, and the sea otter. Baleen whales also can get their baleen plates fouled due to oil spills, which can lead to disaster in feeding areas. The toothed whales, however, use a shiny surfaced and oil repellent skin over a thick blubber layer to thermoregulate, and can move to avoid oil on the surface while feeding on fishes and squid well below the surface. An oil spill in 1990 (the *Mega Borg* spill) off Galveston, Texas, however, showed that bottlenose dolphins do not know how to avoid extensive oil covered areas. Observations revealed that the dolphins repeatedly surfaced in even the very volatile fresh areas of a spill, areas in which humans become sick to their stomachs in minutes and lung tissue becomes rapidly coated in oil volatiles. Under such conditions, the breathing apparatus becomes useless as volatiles are taken into the system, bind with hemoglobin, and effectively suffocate the mammal from a lack of oxygen. Since offshore currents did not bring carcasses close to shore, it is not known how many dolphins succumbed to the *Mega Borg* spill in offshore waters. It can be surmised, however, that fatalities occurred, which emphasizes the need for behavioral studies to determine the fate of toothed whales in oil. This is especially urgent in the Gulf of Mexico, where there is some growing economic pressure to develop oil platforms in those deeper waters frequented by sperm whales and deepwater dolphins.

There are devastating global and point source problems of pollution and other forms of habitat degradation, and the Gulf of Mexico is no exception. The Gulf, however, is not the most badly polluted, giving cause for optimism although not complacency. Dolphins and whales of the Gulf tend to occur in the uppermost strands of complicated trophic webs, and it is necessary to understand these webs and the fates of animals much better.

Concerning problems of increasing habitat degradation, what is to be done? It is unlikely that much of the habitat alteration or outright destruction can be reversed. Pollution, however, can and must be monitored, and its effects are in many cases reversible. Marine mammals offer much promise in helping detect potentially chronic problems in the environment. For example, bottlenose dolphins are long-lived mammals that can be individually identified and tracked in many nearshore areas. If these dolphins are monitored in life and if tissue samples are obtained from dead animals that have beached, then it becomes possible to correlate stored body contaminants with habitat and obtain a better indication about the sources of mortality. Marine mammals are beautiful members of one of our greatest treasures, the world's ocean; we must learn to use their presence to help detect some of the deleterious effects humans are having on our environments. This will help us to enjoy them even more, as well as increase our respect for them.

Another positive development is on the horizon as we head into the twenty-first century. Marine mammals are rapidly becoming subjects of tourism (often called "ecotourism" because of its supposed environmentally friendly nature), not only in the United States and Canada, but worldwide. Tourism based on marine mammals has its recent roots in watching for gray whales from Pacific Coast cliffs, trekking to the elephant seal colony of Año Nuevo, California, boating with killer whales in Puget Sound, following humpback and right whales as they feed off the shores of New England, and, perhaps best known, watching the humpback whales that faithfully reappear to mate and calve off the Hawaiian Islands every winter. Such tourism, when properly carried out, brings money to local economies and is therefore a sustainable resource. It is much better to have 100,000 people see the slowly increasing, yet still decimated, right whales of this world than to slaughter them for the short-term gain of their oil and meat.

Although whale, dolphin, and other marine mammal watching had its roots in the United States during the 1950s, 65 countries and 295 communities had become involved by 1994. The annual revenues of whale watching are conservatively estimated at U.S.$504 million with an annual increase of about 17 percent (1994 statistics). At least twenty towns and cities have been completely transformed by whale and dolphin watching, usually from fishing activities to this sustainable tourism-based economy. Paradoxically, thousands of people come from those countries where marine mammal killing is still taking place (Japan) to see whales and dolphins; but there, as elsewhere, societal mores are rapidly changing, especially with young people who prefer seeing these magnificent creatures on their coastline to having their charred remains on the dinner table. Even as we write, cultural transformations are taking place; in a few years, wholesale slaughter of many (but not all) marine mammals will be a thing of the past.

Properly operated tourism associated with marine mammals can help educate the public, not just about the animals but also general and specific issues of environmental degradation. When a bottlenose dolphin is found with orange–white splotched lesions all over its body, an understanding that this potential pathology might relate to the chemical works farther up the bay could energize people into action and help effect government mandated research, legislation, and enforcement to change these conditions.

In the Gulf of Mexico, tours to see nearshore bottlenose dolphins and manatees have been taking place for the past ten years or so, and are increasing in numbers and locations. Most other marine mammals, however, are far out at sea and therefore necessitate larger vessels to transport people to see them. Nevertheless, such tourism trips, generally combined with marine bird watching, are starting to take hold even in offshore areas of the Gulf of Mexico. Tourism presents its own problems and must be regulated judiciously. For example, it is unwise (and illegal in the United States) to attempt feeding dolphins or manatees; not only could the food be contaminated, but such feeding also makes "pets" of inherently wild creatures, and can keep them from finding food efficiently on their own. Suffice it to say that our love for these creatures has the capacity of being turned into a positive conservation force, not for just them but also for their fragile environments worldwide.

*1.* Eubalaena glacialis, *northern right whale. Painting by Larry Foster*

*2.* Balaenoptera musculus, *blue whale. Painting by Larry Foster*

*3.* Balaenoptera physalus, *fin whale. Painting by Larry Foster*

*4.* Balaenoptera borealis, *Sei whale. Painting by Larry Foster*

5. Balaenoptera edeni, *Bryde's whale. Painting by Larry Foster*

*6.* Balaenoptera acutorostrata, *Minke whale. Painting by Larry Foster*

7. Megaptera novaeangliae, *humpback whale. Painting by Larry Foster*

*8.* Physeter macrocephalus, *sperm whale. Painting by Larry Foster*

*9.* Kogia breviceps, *pygmy sperm whale. Painting by Larry Foster*

*10.* Kogia simus, *dwarf sperm whale. Painting by Larry Foster*

11. Ziphius cavirostris, *Cuvier's beaked whale. Painting by Larry Foster*

*12.* Mesoplodon densirostris, *Blainville's beaked whale. Painting by Larry Foster*

*13.* Mesoplodon bidens, *Sowerby's beaked whale. Painting by Larry Foster*

*14.* Mesoplodon europaeus, *Gervais' beaked whale. Painting by Larry Foster*

*15.* Orcinus orca,
*killer whale.*
*Painting by*
*Larry Foster*

*16.* Globicephala macrorhynchus, *short-finned pilot whale. Painting by Larry Foster*

*17.* Globicephala melas, *long-finned pilot whale. Painting by Larry Foster*

*18.* Pseudorca crassidens, *false killer whale. Painting by Larry Foster*

*19.* Feresa attenuata, *pygmy killer whale. Painting by Larry Foster*

*20.* Peponocephala electra, *melon-headed whale. Painting by Larry Foster*

*21.* Steno bredanensis, *rough-toothed dolphin. Painting by Larry Foster*

*22.* Grampus griseus, *Risso's dolphin. Painting by Larry Foster*

*23.* Tursiops truncatus, *bottlenose dolphin. Painting by Larry Foster*

24. Stenella attenuata, *pantropical spotted dolphin. Painting by Larry Foster*

25. Stenella frontalis, *Atlantic spotted dolphin. Painting by Larry Foster*

*26.* Stenalla longirostris, *spinner dolphin. Painting by Larry Foster*

27. Stenella clymene, *Clymene dolphin. Painting by Larry Foster*

*28.* Stenella coeruleoalba, *striped dolphin. Painting by Larry Foster*

*29.* Delphinus delphis, *short-beaked common dolphin. Painting by Larry Foster*

*30.* Delphinus capensis, *long-beaked common dolphin. Painting by Larry Foster*

*31.* Lagenodelphis hosei, *Fraser's dolphin. Painting by Larry Foster*

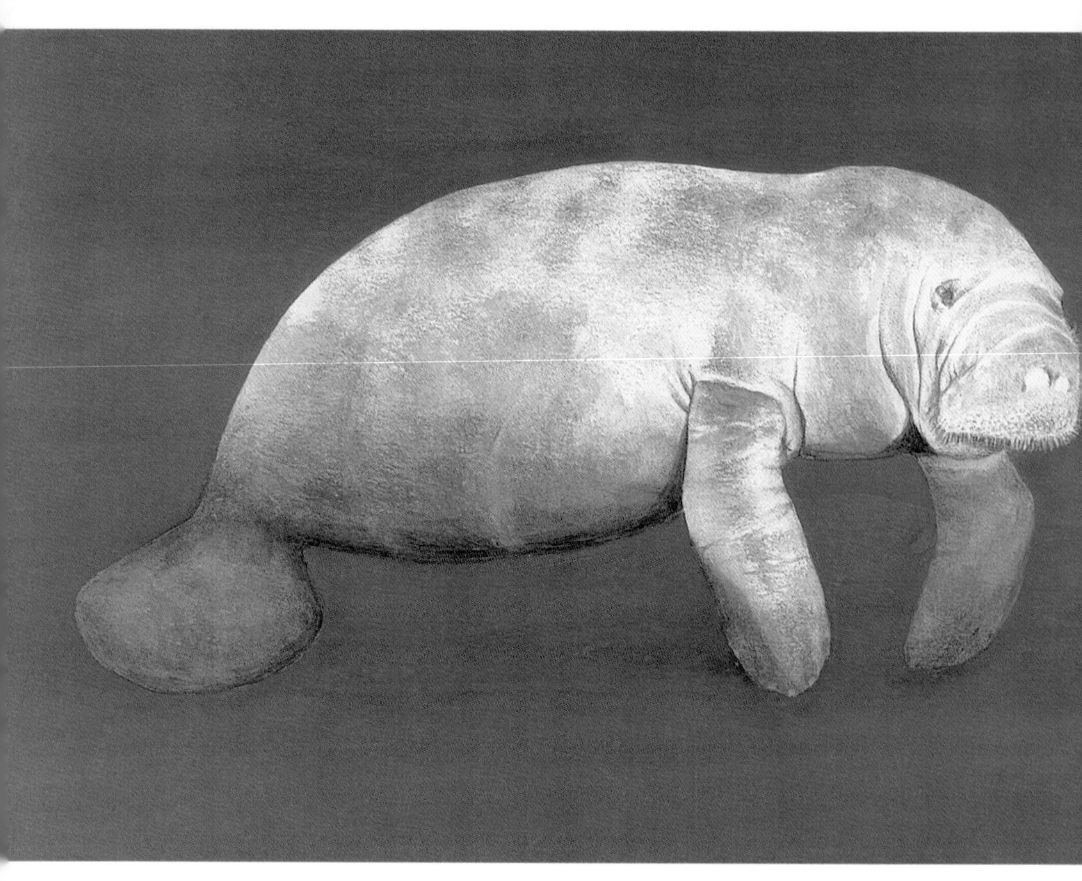

*32.* Trichechus manatus, *West Indian manatee. Painting by Jenny Markowitz*

# Appendix

## *Marine Mammals of the World*

*Note: Those of the Gulf of Mexico indicated by an asterisk (*)*

Order Cetacea (whales, dolphins, and porpoises)

Suborder Mysticeti (baleen whales)

Family Balaenidae (right and bowhead whales)

| | |
|---|---|
| Northern right whale | *Eubalaena glacialis** |
| Southern right whale | *Eubalaena australis* |
| Bowhead whale | *Balaena mysticetus* |

Family Neobalaenidae (pygmy right whale)

| | |
|---|---|
| Pygmy right whale | *Caperea marginata* |

Family Balaenopteridae (The first five of these generally are termed "rorquals.")

| | |
|---|---|
| Blue whale | *Balaenoptera musculus** |
| Fin whale | *Balaenoptera physalus** |
| Sei whale | *Balaenoptera borealis** |
| Bryde's whale | *Balaenoptera edeni** |
| Minke whale | *Balaenoptera acutorostrata** |
| Humpback whale | *Megaptera novaeangliae** |

Family Eschrichtiidae (gray whale)

| | |
|---|---|
| Gray whale | *Eschrichtius robustus* |

Suborder Odontoceti (toothed whales)

Family Physeteridae (sperm whale)

| | |
|---|---|
| Sperm whale | *Physeter macrocephalus** |

Family Kogiidae (pygmy and dwarf sperm whales)

| | |
|---|---|
| Pygmy sperm whale | *Kogia breviceps** |
| Dwarf sperm whale | *Kogia simus** |

Family Monodontidae (narwhal and beluga)

| | |
|---|---|
| Narwhal | *Monodon monoceros* |
| White whale or beluga | *Delphinapterus leucas* |

Family Ziphiidae (beaked whales)

| | |
|---|---|
| Baird's beaked whale | *Berardius bairdii* |
| Arnous's beaked whale | *Berardius arnuxii* |
| Cuvier's beaked whale | *Ziphius cavirostris** |
| Northern bottlenose whale | *Hyperoodon ampullatus* |

| | |
|---|---|
| Southern bottlenose whale | *Hyperoodon planifrons* |
| Shepherd's beaked whale | *Tasmacetus shepherdi* |
| Blainville's beaked whale | *Mesoplodon densirostris** |
| Gray's beaked whale | *Mesoplodon grayi* |
| Ginkgo-toothed beaked whale | *Mesoplodon ginkgodens* |
| Hector's beaked whale | *Mesoplodon hectori* |
| Hubbs' beaked whale | *Mesoplodon carlhubbsi* |
| Pygmy beaked whale | *Mesoplodon peruvianus* |
| Sowerby's beaked whale | *Mesoplodon bidens** |
| Gervais' beaked whale | *Mesoplodon europaeus** |
| True's beaked whale | *Mesoplodon mirus* |
| Strap-toothed whale | *Mesoplodon layardii* |
| Andrew's beaked whale | *Mesoplodon bowdoini* |
| Longman's beaked whale | *Mesoplodon pacificus* |
| Stejneger's beaked whale | *Mesoplodon stejnegeri* |
| Family Delphinidae (ocean dolphins) | |
| Irawaddy dolphin | *Orcaella brevirostris* |
| Killer whale or orca | *Orcinus orca** |
| Short-finned pilot whale | *Globicephala macrorhynchus** |
| Long-finned pilot whale | *Globicephala melas** (possibly in Gulf) |
| False killer whale | *Pseudorca crassidens** |
| Pygmy killer whale | *Feresa attenuata** |
| Melon-headed whale | *Peponocephala electra** |
| Tucuxi | *Sotalia fluviatilis* |
| Indo-Pacific hump-backed dolphin | *Sousa chinensis* |
| Atlantic humpback dolphin | *Sousa teuszii* |
| Rough-toothed dolphin | *Steno bredanensis** |
| Pacific white-sided dolphin | *Lagenorhynchus obliquidens* |
| Dusky dolphin | *Lagenorhynchus obscurus* |
| White-beaked dolphin | *Lagenorhynchus albirostris* |

| | |
|---|---|
| Atlantic white-sided dolphin | *Lagenorhynchus acutus* |
| Hourglass dolphin | *Lagenorhynchus cruciger* |
| Peale's dolphin | *Lagenorhynchus australis* |
| Risso's dolphin | *Grampus griseus** |
| Bottlenose dolphin | *Tursiops truncatus** |
| Pantropical spotted dolphin | *Stenella attenuata** |
| Atlantic spotted dolphin | *Stenella frontalis** |
| Spinner dolphin | *Stenella longirostris** |
| Clymene dolphin | *Stenella clymene** |
| Striped dolphin | *Stenella coeruleoalba** |
| Short-beaked common dolphin | *Delphinus delphis** (possibly inGulf) |
| Long-beaked common dolphin | *Delphinus capensis** (possibly in Gulf) |
| Fraser's dolphin | *Lagenodelphis hosei** |
| Northern right whale dolphin | *Lissodelphis borealis* |
| Southern right whale dolphin | *Lissodelphis peronii* |
| Commerson's dolphin | *Cephalorhynchus commersonii* |
| Heaviside's dolphin | *Cephalorhynchus heavisidii* |
| Hector's dolphin | *Cephalorhynchus hectori* |
| Chilean dolphin | *Cephalorhynchus eutropia* |
| Family Phocoenidae (porpoises) | |
| Dall's porpoise | *Phocoenoides dalli* |
| Spectacled porpoise | *Australophocaena dioptrica* |
| Harbor porpoise | *Phocoena phocoena* |
| Burmeister's porpoise | *Phocoena spinipinnis* |
| Vaquita or Gulf of California harbor porpoise | *Phocoena sinus* |
| Finless porpoise | *Neophocaena phocaenoides* |

| | |
|---|---|
| Family Platanistidae (susu dolphins) | |
| Ganges susu or Ganges River dolphin | *Platanista gangetica* |
| Indus susu or Indus River dolphin | *Platanista minor* |
| Family Iniidae (Boto) | |
| Boto or Amazon River dolphin | *Inia geoffrensis* |
| Family Pontoporiidae (Baiji and Franciscana) | |
| Baiji or Yangtze River dolphin | *Lipotes vexillifer* |
| Franciscana | *Pontoporia blainvillei* |
| Order Sirenia (sea cows) | |
| Family Trichechidae (manatees) | |
| West Indian manatee | *Trichechus manatus** |
| Amazon manatee | *Trichechus inunguis* |
| West African manatee | *Trichechus senegalensis* |
| Family Dugongidae (sea cows and the dugong) | |
| Dugong | *Dugong dugon* |
| Steller's sea cow | *Hydrodamalis gigas* (extinct) |
| Order Carnivora (carnivores) | |
| Suborder Pinnipedia (sea lions, walrus, and seals) | |
| Family Otariidae (fur seals and sea lions) | |
| Steller's sea lion | *Eumetopias jubatus* |
| California sea lion | *Zalophus californianus** (formerly reported from Gulf as escapees) |
| Southern sea lion | *Otaria byronia* |
| Australian sea lion | *Neophoca cinerea* |
| Hooker's sea lion | *Phocarctos hookeri* |
| Northern fur seal | *Callorhinus ursinus* |
| Guadalupe fur seal | *Arctocephalus townsendi* |
| Juan Fernandez fur seal | *Arctocephalus phillippi* |
| Galapagos fur seal | *Arctocephalus galapagoensis* |
| South American fur seal | *Arctocephalus australis* |

| | |
|---|---|
| New Zealand fur seal | *Arctocephalus forsteri* |
| Subantarctic fur seal | *Arctocephalus tropicalis* |
| Antarctic fur seal | *Arctocephalus gazella* |
| South African and Australian fur seal | *Arctocephalus pusillus* |
| Family Odobenidae (walrus) | |
| Walrus | *Odobenus rosmarus* |
| Family Phocidae (true seals) | |
| Harbor seal | *Phoca vitulina* |
| Larga seal | *Phoca largha* |
| Ringed seal | *Phoca hispida* |
| Baikal seal | *Phoca sibirica* |
| Caspian seal | *Phoca caspica* |
| Harp seal | *Phoca groenlandica* |
| Ribbon seal | *Phoca fasciata* |
| Gray seal | *Halichoerus grypus* |
| Bearded seal | *Erignathus barbatus* |
| Hooded seal | *Cystophora cristata* |
| Mediterranean monk seal | *Monachus monachus* |
| West Indian monk seal | *Monachus tropicalis** (formerly in Gulf; now extinct) |
| Hawaiian monk seal | *Monachus schauinslandi* |
| Northern elephant seal | *Mirounga angustirostris* |
| Southern elephant seal | *Mirounga leonina* |
| Crabeater seal | *Lobodon carcinophagus* |
| Ross seal | *Ommatophoca rossii* |
| Leopard seal | *Hydrurga leptonyx* |
| Weddell seal | *Leptonychotes weddellii* |
| Suborder Fissipedia (otters, weasels, minks, and bears) | |
| Family Mustelidae (otters, weasels, and minks) | |
| Sea otter | *Enhydra lutris* |
| Chungungo or marine otter | *Lutra felina* |
| Family Ursidae (bears) | |
| Polar bear | *Ursus maritimus* |

# Bibliography

This section is divided into: (1) a bibliography of general books and easily accessed papers on marine mammals, with special references to animals that occur in the Gulf of Mexico; and (2) a bibliography specific to Gulf marine mammals. The latter has technical references that are more limited in circulation. While many of the general references can be obtained through a bookstore or found in the science section of a good library, the specific references are largely available only through university libraries.

## General References


Andersen, H. T. 1969. *The biology of marine mammals.* New York: Academic Press. 511 pp.

Au, W. W. L. 1993. *The sonar of dolphins.* New York: Springer-Verlag. 277 pp.

Bonner, W. N. 1990. *The natural history of seals.* New York: Facts on File. 196 pp.

Brownell, R. L., Jr., P. B. Best, and J. H. Prescott, eds. 1986. Right whales: Past and present status. Proceedings of the Workshop on the Status of Right Whales. *Rep. Int. Whaling Comm.* Special Issue 10:1–289.

Bryden, M. M., and R. Harrison, eds. 1986. *Research on dolphins.* New York: Oxford University Press. 478 pp.

Calambokidis, J., and G. Steiger. 1997. *Blue whales.* Stillwater, Minn.: Voyageur Press. 72 pp.

Caldwell, D. K., and M. C. Caldwell. 1972. *The world of the bottlenosed dolphin.* Philadelphia: J. B. Lippincott Co. 157 pp.

Clapham, P. 1996. *Humpback whales.* Stillwater, Minn.: Voyageur Press. 72 pp.

Clarke, M. R. 1986. Cephalopods in the diet of odontocetes. In *Research on dolphins,* ed. M. M. Bryden and R. Harrison, 281–321. New York: Oxford University Press.

Connor, R. C., and P. M. Peterson. 1994. *The lives of whales and dolphins.* New York: Henry Holt and Co. 233 pp.

Currey, D., R. Reeve, A. Thornton, and P. Whiting. 1991. *The global war against small cetaceans: A second report by the Environmental Investigation Agency.* London: Environmental Investigation Agency. 63 pp.

Darling, J. D., C. Nicklin, K. S. Norris, H. Whitehead, and B. Würsig. 1995. *Whales, dolphins, and porpoises.* Washington, D.C.: National Geographic Society. 232 pp.

Dierauf, L. A., ed. 1990. *The CRC handbook of marine mammal medicine: Health, disease, and rehabilitation.* Boca Raton: CRC Press. 735 pp.

Domning, D. P. 1996. Bibliography and index of the Sirenia and Desmostylia.

*Smithsonian contributions to paleobiology* no. 80. Washington, D.C.: Smithsonian Institution Press. 611 pp.

Donoghue, M., and A. Wheeler. 1990. *Save the dolphins.* Auckland: David Bateman. 119 pp.

Donovan, G. P. 1980. Sperm whales: Special issue. *Rep. Int. Whaling Comm.* Special Issue 2:1–275.

———. 1982. Aboriginal/subsistence whaling, with special reference to the Alaska and Greenland fisheries. *Rep. Int. Whaling Comm.* 4:1–86.

Donovan, G. P., C. H. Lockyer, and A. L. Martin. 1983. Biology of the northern hemisphere pilot whales. *Rep. Int. Whaling Comm.* Special Issue 14:1–479.

Dudzinski, K., C. W. Clark, and B. Würsig. 1995. A mobile video/acoustic system for simultaneous underwater recording of dolphin interactions. *Aquat. Mammals* 21:187–94.

Ellis, R. 1991. *Men and whales.* New York: Alfred A. Knopf. 542 pp.

Evans, P. G. 1987. *The natural history of whales and dolphins.* New York: Facts on File. 343 pp.

———. 1990. *Whales.* London: Whittet Books. 135 pp.

Gaskin, D. E. 1982. *The ecology of whales and dolphins.* London: Heinemann Educational Books. 459 pp.

Gaskin, D. E., and D. J. St. Aubin, eds. 1990. *Sea mammals and oil: Confronting the risks.* San Diego: Academic Press.

Geraci, J. R., and V. J. Lounsbury. 1993. *Marine mammals ashore: A field guide for strandings.* College Station: Texas A&M University Sea Grant College Program. 305 pp.

Hammond, P. S., S. A. Mizroch, and G. P. Donovan, eds. 1990. Individual recognition of cetaceans: The use of photo-identification and other techniques to estimate population parameters. *Rep. Int. Whaling Comm.* Special Issue 12:1–437.

Harrison, R., and M. M. Bryden, eds. 1988. *Whales, dolphins, and porpoises.* New York: Facts on File. 240 pp.

Hartman, D. S. 1979. *Ecology and behavior of the manatee (Trichechus manatus) in Florida.* Special publication no. 5. Lawrence, Kans.: American Society of Mammalogists.

Herman, L. M., ed. 1980. *Cetacean behavior: Mechanisms and function.* New York: John Wiley & Sons. 463 pp.

Hoelzel, A. R., ed. 1991. Genetic ecology of whales and dolphins. *Rep. Int. Whaling Comm.* Special Issue 13. 311 pp.

Hoese, H. D., and R. H. Moore. 1977. *Fishes of the Gulf of Mexico: Texas, Louisiana, and adjacent waters.* College Station: Texas A&M University Press.

Hofman, R. J. 1995. The changing focus of marine mammal conservation. *Trends Ecol. Evol.* 10:462–65.

Holt, S., and N. M. Young. 1991. *Guide to review of the management of whaling.* Washington, D.C.: Center for Marine Conservation.

Horwood, J. 1987. *The sei whale: Population biology, ecology, and management.* New York: Croom Helm. 375 pp.

Howard, C. J. 1995. *Dolphin chronicles.* New York: Bantam Books. 304 pp.

Hoyt, E. 1995. *The worldwide value and extent of whalewatching 1995.* Bath, England: Whale and Dolphin Conservation Society. 36 pp.

Huntley, A. C., D. P. Costa, G. A. J. Worthy, and M. A. Castellini, eds. 1987. *Marine mammal energetics.* Lawrence, Kans.: Society for Marine Mammalogy.

Jefferson, T. A., S. Leatherwood, and M. A. Webber. 1993. *Marine mammals of the world.* Rome: Food and Agriculture Organization of the United Nations. 320 pp.

Jefferson, T. A., B. Würsig, and D. Fertl. 1993. Cetacean detection and responses to fishing gear. In *Marine mammal sensory systems,* ed. J. A. Thomas, R. A. Kastelein, and A. Ya. Supin, 663–84. New York: Plenum Press.

Kannan, K., and S. Tanabe. 1997. Elevated accumulation of tributyltin and its breakdown products in bottlenose dolphins (*Tursiops truncatus*) found stranded along the U.S. Atlantic and Gulf Coasts—Response. *Environ. Sci. Tech.* 31:3035–36.

Kellogg, W. N. 1961. *Porpoises and sonar.* Chicago: University of Chicago Press. 177 pp.

King, J. E. 1983. *Seals of the world.* Ithaca: Cornell University Press. 240 pp.

Kirk, A. G., and K. G. Vanderhyde, compilers. 1996. *Marine Mammal Commission working bibliography on contaminants in the marine environment and effects on marine mammals,* special report. Washington, D.C.: Marine Mammal Commission.

Kirkevold, B. C., and J. S. Lockard. 1986. Behavioral biology of killer whales. *Zoo biology monographs,* vol. 1. New York: Alan R. Liss. 457 pp.

Klinowska, M., ed. 1991. *Dolphins, porpoises, and whales of the world: The IUCN red data book.* Gland, Switzerland: International Union for the Conservation of Nature. 429 pp.

Laist, D. W. 1997. Impacts of marine debris: Entanglement of marine life in marine debris including a comprehensive list of species with entanglement and ingestion records. In *Marine debris: Sources, impacts, and solutions,* ed. J. M. Coe, 99–413. New York: Springer-Verlag.

Leatherwood, S., and R. R. Reeves. 1983. *The Sierra Club handbook of whales and dolphins.* San Francisco: Sierra Club Books. 302 pp.

———. 1990. *The bottlenose dolphin.* San Diego: Academic Press. 653 pp.

Loughlin, T. R., ed. 1994. *Marine mammals and the Exxon Valdez.* San Diego: Academic Press.

MacDonald, D., ed. 1966. *Whales, dolphins, and porpoises.* Berkeley: University of California Press. 789 pp.

———. 1984. *The encyclopedia of mammals.* New York: Facts on File. 895 pp.

Norris, K. S. 1974. *The porpoise watcher.* New York: W. W. Norton and Co. 250 pp.

———. 1991. *Dolphin days.* New York: W. W. Norton and Co. 335 pp.

Norris, K. S., B. Würsig, R. S. Wells, and M. Würsig. 1994. *The Hawaiian spinner dolphin.* Berkeley: University of California Press. 408 pp.

Norse, E. A., ed. 1993. *Global marine biological diversity: A strategy for building conservation into decision making.* Washington, D.C.: Island Press.

O'Shea, T. J., B. B. Ackerman, and H. F. Percival, eds. 1995. *Population biology of the Florida manatee.* Information and Technology Report 1. Washington, D.C.: U.S. Fish and Wildlife Service. 289 pp.

Payne, R. S., ed. 1983. *Communication and behavior of whales.* Boulder, Colo.: Westview Press. 643 pp.

Payne, R. S., and S. McVay. 1971. Songs of humpback whales. *Science* 173:587–97.

Perrin, W. F. 1975. *Variation of spotted and spinner porpoise (genus Stenella) in the eastern tropical Pacific and Hawaii.* Berkeley: University of California Press. 206 pp.

Pivorunas, A. 1979. The feeding mechanisms of baleen whales. *Am. Sci.* 67: 432–40.

Pryor, K. 1975. *Lads before the wind: Adventures in porpoise training.* New York: Harper and Row. 278 pp.

———. 1995. *On behavior: Essays and research.* North Bend, Wash.: Sunshine Books. 405 pp.

Pryor, K., and K. S. Norris, eds. 1991. *Dolphin societies: Discoveries and puzzles.* Berkeley: University of California Press. 397 pp.

Reeves, R. R., and S. Leatherwood. 1994. *Dolphins, porpoises, and whales: 1994–1998 action plan for the conservation of cetaceans.* Gland, Switzerland: International Union for the Conservation of Nature. 92 pp.

Reeves, R. R., B. S. Stewart, and S. Leatherwood. 1992. *The Sierra Club handbook of seals and sirenians.* San Francisco: Sierra Club Books. 359 pp.

Reijnders, P., S. Brasseur, J. van der Toorn, P. van der Wolf, I. Boyd, J. Harwood, D. Lavigne, and L. Lowry. 1993. *Seals, fur seals, sea lions, and walrus: Status survey and conservation action plan.* Gland, Switzerland: International Union for the Conservation of Nature. 88 pp.

Reynolds, J. E., and D. K. Odell. 1991. *Manatees and dugongs.* New York: Facts on File. 192 pp.

Richardson, W. J., C. R. Greene Jr., C. I. Malme, and D. H. Thomson. 1995. *Marine mammals and noise.* San Diego: Academic Press. 576 pp.

Ridgway, S. H., ed. 1972. *Mammals of the sea: Biology and medicine.* Springfield, Ill.: Charles C. Thomas. 812 pp.

———. 1987. *The dolphin doctor.* New York: Fawcett Crest. 195 pp.

Ridgway, S. H., and R. J. Harrison, eds. 1981. *Handbook of marine mammals.* Vol. 1: *The walrus, sea lions, fur seals, and sea otter.* London: Academic Press. 235 pp.

———, eds. 1981. *Handbook of marine mammals.* Vol. 2: *Seals.* London: Academic Press. 359 pp.

———, eds. 1985. *Handbook of marine mammals.* Vol. 3: *The sirenians and baleen whales.* London: Academic Press. 362 pp.

———, eds. 1989. *Handbook of marine mammals.* Vol. 4: *River dolphins and the larger toothed whales.* London: Academic Press. 442 pp.

———, eds. 1994. *Handbook of marine mammals.* Vol. 5: *The first book of dolphins.* London: Academic Press. 414 pp.

Riedman, M. 1990. *The pinnipeds: Seals, sea lions, and walruses.* Berkeley: University of California Press. 439 pp.

Scammon, C. M. 1968. *The marine mammals of the northwestern coast of North America.* New York: Dover Publications. 319 pp.

Scheffer, V. B. 1976. *A natural history of marine mammals.* New York: Charles Scribner's Sons. 157 pp.

Schevill, W. E. 1962. Whale music. *Oceanus* 9:2–13.

Schusterman, R. J., J. A. Thomas, and F. G. Wood. 1986. *Dolphin cognition and behavior: A comparative approach.* Hillsdale, N.J.: Lawrence Erlbaum Associates. 393 pp.

Shane, S. H. 1988. *The bottlenose dolphin in the wild.* San Carlos, Calif.: Hatcher Trade Press. 48 pp.

Simmonds, M. P., and J. D. Hutchinson, eds. 1996. *The conservation of whales and dolphins: Science and practice.* New York: John Wiley & Sons.

Slijper, E. J. 1976. *Whales and dolphins.* Ann Arbor: University of Michigan Press. 170 pp.

Thompson, P., and B. Wilson. 1994. *Bottlenose dolphins.* Stillwater, Minn.: Voyageur Press. 72 pp.

Tinker, S. W. 1988. *Whales of the world.* Honolulu: Best Press. 310 pp.

Unites States Fish and Wildlife Service. 1995. *Florida manatee recovery plan,* 2d revision. Atlanta: U.S. Fish and Wildlife Service. 160 pp.

Valiela, I. 1995. *Marine ecological processes.* New York: Springer-Verlag.

Watkins, W. A., and D. Wartzok. 1985. Sensory biophysics of marine mammals. *Mar. Mammal Sci.* 1:219–60.

Weller, D. W., V. G. Cockcroft, B. Würsig, S. K. Lynn, and D. Fertl. 1997. Behavioral responses of bottlenose dolphins to remote biopsy sampling and observations of surgical biopsy wound healing. *Aquat. Mammals* 23: 49–58.

Winn, H. K., and B. L. Olla. 1979. *Behavior of marine animals: Current perspectives in research.* Vol. 3: *Cetaceans.* New York: Plenum Press. 438 pp.

Winn, L. K., and H. E. Winn. 1985. *Wings in the sea: The humpback whale.* Hanover, Mass.: University Press of New England. 151 pp.

Worthy, G. A. J. 1990. Nutritional energetics for marine mammals. In *CRC handbook of marine mammal medicine: Health, disease, and rehabilitation,* ed. L. A. Dierauf, 489–520. Boca Raton: CRC Press.

Würsig, B. 1989. Cetaceans. *Science* 244:1550–57.

———. 1990. Cetaceans and oil: Ecological perspectives. In *Sea mammals and oil: Confronting the risks,* ed. J. R. Geraci and D. J. St. Aubin, 129–65. San Diego: Academic Press.

Würsig, B., and T. A. Jefferson. 1990. Methods of photo-identification for small cetaceans. *Rep. Int. Whaling Comm.* Special Issue 12:43–52.

Würsig, B., F. Cipriano, and M. Würsig. 1991. Dolphin movement patterns: Information from radio and theodolite tracking studies. In *Dolphin societies: Discoveries and puzzles,* ed. K. Pryor and K. S. Norris, 79–111. Berkeley: University of California Press.

Würsig, B., T. R. Kieckhefer, and T. A. Jefferson. 1990. Visual displays for communication in cetaceans. In *Sensory abilities of cetaceans,* ed. J. A. Thomas and R. Kastelein, 545–59. New York: Plenum Press.

Young, N. M., S. Iudicello, K. Evans, and D. Baur. 1993. *The incidental capture of marine mammals in U.S. fisheries: Problems and solutions.* Washington, D.C.: Center for Marine Conservation.

## Specific Gulf of Mexico References



Aguayo, C. G. 1954. Notas sobre cetáceos de aguas Cubanas. *Circ. Mus. Biblioteca Zool. Habana* 13:1125–26.

Aguayo L., A., J. P. Gallo R., J. Urbán R., M. Salinas-Z., O. Vidal, and L. T. Findley. 1988. *Beaked whales in Mexican waters.* Scientific Committee Report. Guaymas, México: Universidad de Guaymas.

Barham, E. G., J. C. Sweeney, S. Leatherwood, R. K. Beggs, and C. L. Barham. 1980. Aerial census of the bottlenose dolphin, *Tursiops truncatus,* in a region of the Texas coast. *Fish. Bull.* 77:585–95.

Barron, G. L., and T. A. Jefferson. 1993. First records of the melon-headed whale (*Peponocephala electra*) from the Gulf of Mexico. *Southwest. Nat.* 38:82–85.

Barros, N. B. 1987. Food habits of bottlenose dolphins (*Tursiops truncatus*) in the southeastern United States, with special reference to Florida waters. M.S. thesis, University of Miami, Miami.

———. 1993. Feeding ecology and foraging strategies of bottlenose dolphins in the central east coast of Florida. Ph.D. dissertation, University of Miami, Coral Gables, Fla.

Barros, N. B., and D. K. Odell. 1990. Food habits of bottlenose dolphins in the southeastern United States. In *The bottlenose dolphin,* ed. S. Leatherwood and R. R. Reeves, 309–28. San Diego: Academic Press.

Baughman, J. L. 1946. Dolphins. *Tex. Game and Fish* 4:11, 20.

———. 1946. On the occurrence of a rorqual whale on the Texas coast. *J. Mammal.* 27:392–93.

Baumgartner, M. F. 1995. The distribution of select species of cetaceans in the northern Gulf of Mexico in relation to observed environmental variables. M.S. thesis, University of Southern Mississippi, Hattiesburg.

———. 1997. The distribution of Risso's dolphin (*Grampus griseus*) with respect to the physiography of the northern Gulf of Mexico. *Mar. Mammal Sci.* 13:614–38.

Biggs, D. C., and R. W. Davis. 1996. Thar she blows!: GulfCet cruise news. *Quarterdeck, Newsl. Dept. Oceanogr.,* Texas A&M University 4(3):18–19.

Blaylock, R. A, J. W. Hain, L. J. Hansen, D. L. Palka, and G. T. Waring. 1995. U.S. Atlantic and Gulf of Mexico marine mammal stock assessments. *Nat. Ocean. Atm. Admin. Tech. Memo., Nat. Mar. Fish. Serv., Southeast Fish. Sci. Ctr.* 363:1–211.

Blaylock, R. A., and W. Hoggard. 1994. Preliminary estimates of bottlenose dolphin abundance in southern U.S. Atlantic and Gulf of Mexico Continental Shelf waters. *Nat. Ocean. Atm. Admin. Tech. Memo., Nat. Mar. Fish. Serv., Southeast Fish. Sci. Ctr.* 356:1–10.

Bonde, R. K., and T. J. O'Shea. 1989. Sowerby's beaked whale (*Mesoplodon bidens*) in the Gulf of Mexico. *J. Mammal.* 70:447–49.

Bonfil, R. 1997. Status of shark resources in the southern Gulf of Mexico and Caribbean: Implications for management. *Fish. Res.* 29:101–17.

Bossart, G. D., D. K. Odell, and N. H. Altman. 1985. Cardiomyopathy in stranded pygmy and dwarf sperm whales. *J. Am. Vet. Med. Assoc.* 187:1137–40.

Bräger, S. 1992. Untersuchungen zur Ortstreue und zum Vergesellschaftungsmuster des Grossen Tümmlers, *Tursiops truncatus* (Montagu, 1821). Diplom, Christian-Albrechts-Universität, Kiel, Germany.

———. 1993. Diurnal and seasonal behavior patterns of bottlenose dolphins (*Tursiops truncatus*). *Mar. Mammal Sci.* 9:434–38.

Bräger, S., B. Würsig, A. Acevedo, and T. Henningsen. 1994. Association patterns of bottlenose dolphins (*Tursiops truncatus*) in Galveston Bay, Texas. *J. Mammal.* 75:431–37.

Breuer, J. P. 1951. Gilchrist's whale. *Tex. Game and Fish* 9:24–25.

Brown, D. H., D. K. Caldwell, and M. C. Caldwell. 1966. Observations on the behavior of wild and captive false killer whales, with notes on associated behavior of other genera of captive dolphins. *L.A. County Mus. Contrib. Sci.* 95:1–32.

Buck, J. D. 1984. Microbiological observations on two stranded live whales. *J. Wildl. Dis.* 20:148–50.

Buck, J. D., and S. A. McCarthy. 1994. Occurrence of non-01 *Vibrio cholerae* in Texas Gulf Coast dolphins (*Tursiops truncatus*). *Lett. Appl. Microbiol.* 18: 45–46.

Bull, A. S., and J. J. Candle. 1994. An indication of the process: Offshore platforms as artificial reefs in the Gulf of Mexico. *Bull. Mar. Sci.* 55:1086–89.

Bullis, H. R., and J. C. Moore. 1956. Two occurrences of false killer whales, and a summary of American records. *Am. Mus.* (November) 1756:1–5.

Burn, D. M., and G. P. Scott. 1988. *Synopsis of available information on marine mammal–fisheries interactions in the southeastern United States: Preliminary report.* Contribution C.D.-87/88-26. Miami: National Marine Fisheries Service, Southeast Fisheries Science Center.

Caldwell, D. K. 1955. Notes on the spotted dolphin, *Stenella plagiodon,* and the first record of the common dolphin, *Delphinus delphis,* in the Gulf of Mexico. *J. Mammal.* 36:467–70.

———. 1960. Notes on the spotted dolphin in the Gulf of Mexico. *J. Mammal.* 41:134–36.

Caldwell, D. K., and M. C. Caldwell. 1966. Observations on the distribution, coloration, behavior, and audible sound production of the spotted dolphin, *Stenella plagiodon* (Cope). *L.A. County Mus. Contrib. Sci.* 104:1–28.

———. 1969. Gray's dolphin, *Stenella styx,* in the Gulf of Mexico. *J. Mammal.* 50:612–14.

———. 1971. Sounds produced by two rare cetaceans stranded in Florida. *Cetology* 4:1–6.

———. 1971. The pygmy killer whale, *Feresa attenuata,* in the western Atlantic, with a summary of world records. *J. Mammal.* 52:206–209.

———. 1971. Underwater pulsed sounds produced by captive spotted dolphins, *Stenella plagiodon. Cetology* 1:1–7.

———. 1973. Marine mammals of the eastern Gulf of Mexico. In *A summary of knowledge of the eastern Gulf of Mexico,* ed. J. I. Jones, R. E. Ring, M. O. Rinkel, and R. E. Smith, III-1:23. Gainesville: State University System of Florida.

———. 1975. Dolphin and small whale fisheries of the Caribbean and West Indies: Occurrence, history, and catch statistics, with special reference to the Lesser Antillian island of St. Vincent. *J. Fish. Res. Board Can.* 32:1105–10.

———. 1975. Pygmy killer whales and short-snouted spinner dolphins in Florida. *Cetology* 18:1–5.

———. 1989. Pygmy sperm whale, *Kogia breviceps* (de Blainville, 1838) and dwarf sperm whale, *Kogia simus* (Owen, 1866). In *Handbook of marine mammals,* vol. 4, ed. S. H. Ridgway and R. Harrison, 235–60. London: Academic Press.

Caldwell, D. K., M. C. Caldwell, and D. W. Rice. 1966. Behavior of the sperm whale, *Physeter catodon* L. In *Whales, dolphins, and porpoises,* ed. K. S. Norris, 678–717. Berkeley: University of California Press.

Caldwell, D. K., M. C. Caldwell, and C. M. Walker Jr. 1970. Mass and individual strandings of false killer whales, *Pseudorca crassidens,* in Florida. *J. Mammal.* 51:634–36.

Caldwell, D. K., J. N. Layne, and J. B. Siebenaler. 1956. Notes on a killer whale (*Orcinus orca*) from the northwestern Gulf of Mexico. *Quart. J. Fla. Acad. Sci.* 19:189–96.

Caldwell, D. K., J. H. Prescott, and M. C. Caldwell. 1966. Production of pulsed sounds by the pygmy sperm whale, *Kogia breviceps. Bull. So. Calif. Acad. Sci.* 65:245–48.

Caldwell, D. K., W. F. Rathjen, and M. C. Caldwell. 1970. Pilot whales mass stranded at Nevis, West Indies. *Quart. J. Fla. Acad. Sci.* 33:241–43.

Caldwell, D. K., J. B. Siebenaler, and A. Inglis. 1960. Sperm and pygmy sperm whales stranded in the Gulf of Mexico. *J. Mammal.* 41:136–38.

Caldwell, D. K., J. B. Siebenaler, J. C. Woodard, L. Ajello, W. Kaplan, and H. M. McClure. 1975. Lobomycosis as a disease of the Atlantic bottlenosed dolphin (*Tursiops truncatus* Montagu, 1821). *Am. J. Trop. Med. Hygiene* 24:105–14.

Caldwell, M. C., D. K. Caldwell, and J. B. Siebenaler. 1965. Observations on captive and wild Atlantic bottlenose dolphins, *Tursiops truncatus,* in the northeastern Gulf of Mexico. *L.A. County Mus. Contrib. Science* 91:1–10.

Campbell, H. W., and D. Gicca. 1978. Reseña preliminar del estado actual y distribución del manatí (*Trichechus manatus*) en México. *Anales Instituto Biologia, Universidad Nacional Autónoma de México, Serie Zoología* 49:257–64.

Clark, A. H. 1884. Mammals. In *The fisheries and fishery industries of the United States,* ed. G. B. Goode, 7–32. Washington, D.C.: U.S. Government Printing Office.

Clarke, M. R. 1976. Observations on sperm whale diving. *J. Mar. Biol. Assoc. U.K.* 56:809–10.

———. 1979. The head of the sperm whale. *Sci. Am.* 240:128–41.

Collum, L. A., and T. H. Fritts. 1985. Sperm whales (*Physeter catodon*) in the Gulf of Mexico. *Southwest. Nat.* 30:101–104.

Cowan, D. F. 1993. Lobo's disease in a bottlenose dolphin (*Tursiops truncatus*) from Matagorda Bay, Texas. *J. Wildl. Dis.* 29:488–89.

Crowder, B. 1980. Whales in the Gulf of Mexico? Of course! *The University and the Sea* 3:6–10.

Cumbaa, S. L. 1980. Aboriginal use of marine mammals in the southeastern United States. *Southeast. Archaeol. Conf. Bull.* 17:6–10.

Cummings, W. C., and P. O. Thompson. 1971. Underwater sounds from the blue whale, *Balaenoptera musculus. J. Acoust. Soc. Am.* 50:1193–98.

Cuni, L. A. 1918. Contribución al estudio de mamíferos acuáticos observados en las costas de Cuba. *Memorias de la Sociedad Cubana de Historia Natural "Felipe Poey"* 3:83–123.

Curry, B. E. 1997. Phylogenetic relationships among bottlenose dolphins (genus: *Tursiops*) in a worldwide context. Ph.D. dissertation, Texas A&M University, College Station.

Darnell, R. M., and R. E. Defenbaugh. 1990. Gulf of Mexico environmental overview and history of environmental research. *Am. Zool.* 30:3–6.

Davis, J. W. 1993. An analysis of tissues for total PCB and planar PCB concentrations in marine mammals stranded along the Gulf of Mexico. M.S. thesis, Texas A&M University, College Station. 144 pp.

Davis, R. W., and G. S. Fargion, eds. 1996. *Distribution and abundance of marine mammals in the north–central and western Gulf of Mexico: Final report.* Vol. I: *Executive summary.* Herndon, Va.: Minerals Management Service.

———, eds. 1996. *Distribution and abundance of marine mammals in the north–central and western Gulf of Mexico: Final report.* Vol. II: *Technical report.* Herndon, Va.: Minerals Management Service.

Davis, R. W., G. A. J. Worthy, B. Würsig, S. K. Lynn, and F. I. Townsend. 1996. Diving behavior and at-sea movements of an Atlantic spotted dolphin in the Gulf of Mexico. *Mar. Mammal Sci.* 12:569–81.

Davis, R. W., G. S. Fargion, N. May, T. D. Leming, M. Baumgartner, W. E. Evans, L. J. Hansen, and K. Mullin. 1998. Physical habitats of cetaceans along the Continental Slope in the northcentral and western Gulf of Mexico. *Mar. Mammal Sci.* 14:490–507.

Davis, R. W., G. Scott, B. Würsig, W. Evans, G. Fargion, L. Hansen, R. Benson, K. Mullin, N. May, T. Leming, B. Mate, J. Norris, and T. Jefferson. 1994. *Distribution and abundance of marine mammals in the north–central and western Gulf of Mexico.* Outer Continental Shelf Study, 94-0003. Herndon, Va.: Minerals Management Service. 129 pp. + appendices.

Davis, W. B., and D. J. Schmidly. 1994. *The mammals of Texas.* Austin: University of Texas Press.

———. 1994. The mammals of Texas. *Tex. Parks Wildl. Dept. Bull.* 41:1–338.

Defenbaugh, R. E. 1990. The Gulf of Mexico—A management perspective. *Am. Zool.* 30:7–13.

de la Osa, J. A., and J. G. Guma. 1971. Pilot whales captured off the coast of Cuba. *Transl. Lat. Am.* 634:78–79.

Delgado-Estrella, A. 1991. Algunos aspectos de la ecología de poblaciones de las toninas (*Tursiops truncatus Montagu,* 1821) en la Laguna de Términos y Sonda de Campeche, México. Tesis, Universidad Nacional Autónoma de México.

———. 1994. Presencia del delfin dientes rugosos o esteno (*Steno bredanensis*), en la costa de Tabasco, México. *Anales Inst. Biól. Univ. Nac. Autón México, Ser. Zool.* 65:303–305.

Domning, D. P., and L. A. C. Layek. 1986. Interspecific and intraspecific morphological variation in manatees (Sirenia: *Trichechus*). *Mar. Mammal Sci.* 2:87–144.

Dorf, B. A. 1982. Oceanographic factors and cetacean distributions at two sites in the Gulf of Mexico. M.S. thesis, Texas A&M University, College Station. 135 pp.

Duffield, D. A. 1986. *Investigations of genetic variability in stocks of the bottlenose dolphin (Tursiops truncatus) and the loggerhead sea turtle (Caretta caretta).* Miami: Final Contract Report National Marine Fisheries Service, Southeast Fisheries Science Center.

Duffield, D. A., and J. Chamberlin-Lea. 1990. Use of chromosome heteromorphisms and hemoglobins in studies of bottlenose dolphin populations and paternities. In *The bottlenose dolphin,* ed. S. Leatherwood and R. R. Reeves, 609–20. San Diego: Academic Press.

Duffield, D. A., and R. S. Wells. 1991. The combined application of chromosome, protein, and molecular data for the investigation of social unit structure and dynamics in *Tursiops truncatus. Rep. Int. Whaling Comm.* Special Issue 13:155–69.

Duignan, P. J., C. House, D. K. Odell, R. S. Wells, L. J. Hansen, M. T. Walsh, D. J. St. Aubin, B. K. Rima, and J. R. Geraci. 1996. Morbillivirus infection in bottlenose dolphins: Evidence for recurrent epizootics in the western Atlantic and Gulf of Mexico. *Mar. Mammal Sci.* 12:499–515.

Duignan, P. J., C. House, M. T. Walsh, T. Campbell, G. D. Bossart, N. Duffy, P. J. Fernandes, B. K. Rima, S. Wright, and J. R. Geraci. 1995. Morbillivirus infection in manatees. *Mar. Mammal Sci.* 11:441–51.

Duncan, C. D., and R. W. Havard. 1980. Pelagic birds of the northern Gulf of Mexico. *Am. Birds* 34:122–32.

Duncan, M. 1997. Analysis of an arrayed acoustic listening system for sperm whale studies. M.S. thesis, Texas A&M University, College Station.

Edds, P. L., D. K. Odell, and B. R. Tershy. 1993. Vocalizations of a captive juvenile and free-ranging adult–calf pair of Bryde's whales, *Balaenoptera edeni. Mar. Mammal Sci.* 9:269–84.

Edwards, J. R. 1995. Dolphin watching. *Tex. Parks Wildl.* 53(6):6–13.

Fehring, W. K., and R. S. Wells. 1976. A series of strandings by a single herd of pilot whales on the west coast of Florida. *J. Mammal.* 57:191–94.

Fernandez, S. P. 1992. Composición de edad y sexo y parámetros del ciclo de vida de Toninas (*Tursiops truncatus*) varadadas en el noroeste del Golfo de México. M.S. tesis, Instituto Tecnológico y de Estudios Superiores de Monterrey, México.

Fernandez, S. P., and S. C. Jones. 1990. Manatee stranding on the coast of Texas. *Tex. J. Sci.* 42:103.

Fertl, D. C. 1994. Occurrence, movements, and behavior of bottlenose dolphins (*Tursiops truncatus*) in association with the shrimp fishery in Galveston Bay, Texas. M.S. thesis, Texas A&M University, College Station.

———. 1994. Occurrence patterns and behavior of bottlenose dolphins (*Tursiops truncatus*) in the Galveston ship channel, Texas. *Tex. J. Sci.* 46(4): 299–317.

Fertl, D. C., and S. Leatherwood. In press. Cetacean interactions with trawls: A preliminary review. *J. Northwest Atlantic Fish. Sci.* 21.

Fertl, D. C., and A. Schiro. 1994. Carrying of dead calves by free-ranging Texas bottlenose dolphins (*Tursiops truncatus*). *Aquat. Mammals* 20:53–56.

Fertl, D. C., and B. Wilson. 1997. Bubble use during prey capture by a lone bottlenose dolphin (*Tursiops truncatus*). *Aquat. Mammals* 23(2):113–14.

Fertl, D. C., and B. Würsig. 1993. Shrimp boats: A Galveston dolphin's smorgasbord. *Galveston Bay Foundation Soundings* 5(2):10–12.

———. 1995. Coordinated feeding by Atlantic spotted dolphins (*Stenella frontalis*) in the Gulf of Mexico. *Aquat. Mammals* 21:3-5.

Fertl, D. C., A. J. Schiro, and D. Peake. 1997. Coordinated feeding by Clymene dolphins (*Stenella clymene*) in the Gulf of Mexico. *Aquat. Mammals* 23:111–12.

Fertl, D. C., B. Würsig, and K. D. Mullin. In press. Exploring new frontiers: The Gulf of Mexico's cetaceans. *Whalewatcher.*

Forrester, D. J., and W. D. Robertson. 1975. Helminths of rough-toothed dolphins, *Steno bredanensis* Lesson, 1828, from Florida waters. *J. Parasitol.* 61:922.

Forrester, D. J., D. K. Odell, N. P. Thompson, and J. R. White. 1980. Morphometrics, parasites, and chlorinated hydrocarbon residues of pygmy killer whales from Florida. *J. Mammal.* 61:356–60.

Fritts, T. H., and R. P. Reynolds. 1981. *Pilot study of the marine mammals, birds, and turtles in the OCS area of the Gulf of Mexico. Final report.* Miami: National Marine Fisheries Service. 139 pp.

Fritts, T. H., W. Hoffman, and M. A. McGehee. 1983. The distribution and abundance of marine turtles in the Gulf of Mexico and nearby Atlantic waters. *J. Herpetol.* 17:327–44.

Fritts, T. H., A. B. Irvine, R. D. Jennings, L. A. Collum, W. Hoffman, and M. A. McGehee. 1983. *Turtles, birds, and mammals in the northern Gulf of Mexico and nearby Atlantic waters. Final report.* Washington, D.C.: U.S. Fish and Wildlife Service. 455 pp.

Gallo-Reynoso, J. P. 1989. Primer registro del zifio de las Antillas (*Mesoplodon europaeus* Gervais, 1955) (Cetacea: Ziphiidae) en México. *Anales Inst. Biol. Univ. Nac. Autón. México, Ser. Zool.* 60:267–78.

Gambell, R. 1976. World whale stocks. *Mammal Rev.* 6:41–53.

Gitschlag, G. R., and B. A. Herczeg. 1994. Sea turtle observations at explosive removals of energy structures. *Mar. Fish. Rev.* 56:1–8.

Goodwin, D. E. 1985. Diurnal behavior patterns of *Tursiops truncatus* off Mobile Point, Alabama. M.S. thesis, San Francisco State University.

Gruber, J. A. 1981. Ecology of the Atlantic bottlenosed dolphin (*Tursiops truncatus*) in the Pass Cavallo area of Matagorda Bay, Texas. M.S. thesis, Texas A&M University, College Station. 200 pp.

Gunter, G. 1941. A record of the long-snouted dolphin, *Stenella plagiodon* (Cope), from the Texas coast. *J. Mammal.* 22:447–48.

———. 1941. Death of fishes due to cold on the Texas coast, January, 1940. *Ecology* 22:203–208.

———. 1941. Occurrence of the manatee in the United States, with records from Texas. *J. Mammal.* 22:60–64.

———. 1942. Contributions to the natural history of the bottlenose dolphin, *Tursiops truncatus* (Montague), on the Texas coast, with particular reference to food habits. *J. Mammal.* 23:267–76.

———. 1943. Letter to the editor. *J. Mammal.* 24:521.

———. 1944. Texas porpoises. *Tex. Game and Fish* 2:11.

———. 1946. Records of the blackfish or pilot whales from the Texas coast. *J. Mammal.* 27:374–77.

———. 1947. Sight records of the West Indian seal, *Monachus tropicalis* (Gray), from the Texas coast. *J. Mammal.* 28:289–90.

———. 1951. Consumption of shrimp by the bottle-nosed dolphin. *J. Mammal.* 32:465–66.

———. 1954. Mammals of the Gulf of Mexico. *Fish. Bull.* 55:543–51.

———. 1955. Blainville's beaked whale, *Mesoplodon densirostris,* on the Texas coast. *J. Mammal.* 36:573–74.

———. 1968. The status of seals in the Gulf of Mexico with a record of feral otariid seals off the United States Gulf Coast. *Gulf Res. Rep.* 2:301–308.

———. 1977. Public aquaria and seals in the Gulf of Mexico. *Drum and Croaker* 17:37–40.

Gunter, G., and J. Y. Christmas. 1973. Stranding record of a finback whale, *Balaenoptera physalus,* from Mississippi and the goose-beaked whale, *Ziphius cavirostris,* from Louisiana. *Gulf Res. Rep.* 4:168–73.

Gunter, G., and G. Corcoran. 1981. Mississippi manatees. *Gulf Res. Rep.* 7:97–99.

Gunter, G., and R. Overstreet. 1974. Cetacean notes. I. Sei and rorqual whales on the Mississippi coast, a correction. II. A dwarf sperm whale in Mississippi Sound and its helminth parasites. *Gulf Res. Rep.* 4:479–81.

Gunter, G., and A. Perry. 1983. A 1981 sighting of *Trichechus manatus* in Mississippi. *J. Mammal.* 64:513.

Gunter, G., C. L. Hubbs, and M. A. Beal. 1955. Records of *Kogia breviceps* from Texas, with remarks on movements and distribution. *J. Mammal.* 36:263–70.

Haebler, R., and A. Hohn, eds. In press. *Dolphins: Factors in morbidity and mortality.* Silver Springs, Md.: NOAA Technical Report.

Hamilton, P. V., and R. T. Nishimoto. 1977. Dolphin predation on mullet. *Fla. Sci.* 40:251–52.

Hamilton, W. J. 1941. Notes on some mammals of Lee County, Florida. *Am. Midl. Nat.* 25:686–91.

Hancock, D. 1965. Killer whales kill and eat a minke whale. *J. Mammal.* 46: 341–42.

Hanlon, R. T. 1985. Cephalopods of the northwestern Gulf of Mexico. *Tex. Conchol.* 21(3):90–93.

Hansen, L. J. (coord.). 1992. *Report on the investigation of 1990 Gulf of Mexico bottlenose dolphin strandings.* Contribution MIA-9293. Miami: NMFS Southeast Fisheries Science Center.

Harris, S. A. 1986. Beached! *La. Conserv.* 38:18–22.

———. 1987. Caught between the devil and the deep blue sea. *La. Conserv.* 39:4–9.

Harrison, R. J., and S. H. Ridgway. 1971. Gonadal activity in some bottlenose dolphins (*Tursiops truncatus*). *J. Zool.* (London) 165:355–66.

Hartman, D. S. 1979. *Ecology and behavior of the manatee (Trichechus manatus) in Florida.* Special publication no. 5. Lawrence, Kans.: American Society of Mammalogists.

Haubold, E. 1995. Selected heavy metals in tissues of bottlenose dolphins (*Tursiops truncatus*) stranded along the Texas and Florida Gulf Coasts. M.S. thesis, Texas A&M University, College Station.

Henningsen, T. 1991. Zur Verbreitung und Ökologie des Grossen Tümmlers (*Tursiops truncatus*) in Galveston, Texas. Diplom, Christian-Albrechts-Universität, Kiel, Germany.

Henningsen, T., and B. Würsig. 1991. Bottlenose dolphins in Galveston Bay, Texas: Numbers and activities. In *Proceedings of the 1991 European Society Conference,* 36–38. Cambridge: Cambridge University Press.

———. 1991. Interactions between humans and dolphins in Galveston Bay, Texas. In *Proceedings of the symposium Whales: Biology, threats, and conservation, Brussels June 5–7, 1991,* ed. J. J. Symoens, 135–40. Brussels: Royal Academy of Overseas Sciences.

Hersh, S. L., and D. A. Duffield. 1990. Distinction between Atlantic offshore and coastal bottlenose dolphins based on hemoglobin profile and morphometry. In *The Bottlenose Dolphin,* ed. S. Leatherwood and R. R. Reeves, 129–39. San Diego: Academic Press.

Hersh, S. L., and D. K. Odell. 1986. Mass stranding of Fraser's dolphin, *Lagenodelphis hosei,* in the western North Atlantic. *Mar. Mammal Sci.* 2: 73–76.

Herzing, D. L. 1997. The life history of free-ranging Atlantic spotted dolphins (*Stenella frontalis*): Age classes, color phases, and female reproduction. *Mar. Mammal Sci.* 13:576–95.

Hoffman, W., T. H. Fritts, and R. P. Reynolds. 1981. Whale sharks associated with fish schools off south Texas. *Northeast Gulf Sci.* 5:55–57.

Hohn, A. A., M. D. Scott, R. S. Wells, A. B. Irvine, and J. C. Sweeney. 1989. Growth layers in teeth from known-age, free-ranging bottlenose dolphins. *Mar. Mammal Sci.* 5:315–42.

Hugentobler, H., and J. P. Gallo. 1986. Un registro de la estenela moteada del

Atlantico *Stenella plagiodon* (Cope, 1866) (Cetacea: Delphinidae) del estado de Campeche, Mexico. *Anales Inst. Biol. Univ. Nac. Autón. México, Ser. Zool.* 56:1039–42.

Hysmith, B. T., R. T. Colura, and J. C. Kana. 1976. Beaching of a live pygmy sperm whale (*Kogia breviceps*) at Port O'Connor, Texas. *Southwest. Nat.* 21:409.

Irvine, A. B., and R. S. Wells. 1972. Results of an attempt to tag Atlantic bottlenosed dolphins, *Tursiops truncatus. Cetology* 13:1–5.

Irvine, A. B., R. S. Wells, and M. D. Scott. 1982. An evaluation of techniques for tagging small odontocete cetaceans. *Fish. Bull.* 80:671–88.

Irvine, A. B., M. D. Scott, R. S. Wells, and J. H. Kaufman. 1981. Movements and activities of the Atlantic bottlenose dolphin, *Tursiops truncatus,* near Sarasota, Florida. *Fish. Bull.* 79:671–88.

James, P., F. W. Judd, and J. C. Moore. 1970. First western Atlantic occurrence of the pygmy killer whale. *Fieldiana Zool.* 58:1–3.

Jefferson, T. A. 1995. Distribution, abundance, and some aspects of the biology of cetaceans in the offshore Gulf of Mexico. Ph.D. dissertation, Texas A&M University, College Station. 232 pp.

———. 1996. Estimates of abundance of cetaceans in offshore waters of the northwestern Gulf of Mexico, 1992–1993. *Southwest. Nat.* 41:279–87.

———. 1996. Morphology of the Clymene dolphin (*Stenella clymene*) in the northern Gulf of Mexico. *Aquat. Mammals* 22:35–43.

Jefferson, T. A., and G. D. Baumgardner. 1997. Osteological specimens of marine mammals (Cetacea and Sirenia) from the western Gulf of Mexico. *Tex. J. Sci.* 49:97–108.

Jefferson, T. A., and S. Leatherwood. 1994. *Lagenodelphis hosei. Mammal. Species* 147:1–5.

Jefferson, T. A., and S. K. Lynn. 1994. Marine mammal sightings in the Gulf of Mexico and Caribbean Sea, Summer 1991. *Caribb. J. Sci.* 30:83–89.

Jefferson, T. A., and K. D. Mullin. 1995. Flipper, Moby Dick, et al. *Tex. Parks Wildl.* 53(7):22–25.

Jefferson, T. A., and A. J. Schiro. 1997. Distribution of cetaceans in the offshore Gulf of Mexico. *Mammal Rev.* 27:27–50.

Jefferson, T. A., D. K. Odell, and K. T. Prunier. 1995. Notes on the biology of the Clymene dolphin (*Stenella clymene*) in the northern Gulf of Mexico. *Mar. Mammal Sci.* 11:564–73.

Jefferson, T. A., S. Leatherwood, L. K. M. Shoda, and R. L. Pitman. 1992. *Marine mammals of the Gulf of Mexico: A field guide for aerial and shipboard observers.* College Station: Texas A&M University Printing Center. 92 pp.

Jennings, R. 1982. Pelagic sightings of Risso's dolphin, *Grampus griseus,* in the Gulf of Mexico and Atlantic Ocean adjacent to Florida. *J. Mammal.* 63: 522–23.

Jones, S. C. 1986. It's happening in Texas. *Whalewatcher* 20:8–11.

———. 1987. Patterns of recent marine mammal strandings along the upper Texas coast. *Cetus* 7:10–15.

———. 1988. Survey of the Atlantic bottlenose dolphin (*Tursiops truncatus*) population near Galveston, Texas. M.S. thesis, Texas A&M University, College Station. 59 pp.

Kannan, K., K. Senthilkumar, B. G. Loganathan, S. Takahashi, D. K. Odell, and S. Tanabe. 1997. Elevated accumulation of tributyltin and its breakdown products in bottlenose dolphins (*Tursiops truncatus*) found stranded along the U.S. Atlantic and Gulf Coasts. *Environ. Sci. Tech.* 31:296–301.

Katona, S. K., J. A. Beard, P. E. Girton, and F. Wenzel. 1988. Killer whales (*Orcinus orca*) from the Bay of Fundy to the equator, including the Gulf of Mexico. *Rit Fiskideildar* 11:205–24.

Klima, E. F., G. R. Gitschlag, and M. L. Renaud. 1988. Impacts of the explosive removal of offshore petroleum platforms on sea turtles and dolphins. *Mar. Fish. Rev.* 59:33–42.

Krafft, A. K., J. H. Lichy, T. P. Lipscomb, B. A. Klaunberg, S. Kennedy, and J. K. Taubenberger. 1995. Postmortem diagnosis of morbillivirus infection in bottlenose dolphins (*Tursiops truncatus*) in the Atlantic and Gulf of Mexico epizootics by polymerase chain reaction-based assay. *J. Wildl. Dis.* 31:410–15.

Kuehl, D. W., and R. Haebler. 1995. Organochlorine, organobromine, metal, and selenium residues in bottlenose dolphins (*Tursiops truncatus*) collected during an unusual mortality event in the Gulf of Mexico, 1990. *Arch. Environ. Contam. Toxicol.* 28:494–99.

Lahvis, G. P., R. S. Wells, D. W. Kuehl, J. L. Stewart, H. L. Rhinehart, and C. S. Via. 1995. Decreased lymphocyte responses in free-ranging bottlenose dolphins (*Tursiops truncatus*) are associated with increased concentrations of PCBs and DDT in peripheral blood. *Environ. Health Pers.* 103:67–72.

Layne, J. N. 1965. Observations on marine mammals in Florida waters. *Bull. Fla. State Mus.* 9:131–81.

Lazcano-Barrero, M. A., and J. M. Packard. 1989. The occurrence of manatees (*Trichechus manatus*) in Tamaulipas, Mexico. *Mar. Mammal Sci.* 5:202–205.

Leahy, T. M. 1977. The mystery of the beached mammals. *Nat. Ocean. Atm. Admin. Mag.* 7:4–8.

Leatherwood, S. 1975. Some observations of feeding behavior of bottlenosed dolphins (*Tursiops truncatus*) in the northern Gulf of Mexico and (*Tursiops* cf. *T. gilli*) off southern California, Baja California, and Nayarit, Mexico. *Mar. Fish. Rev.* 37:10–16.

Leatherwood, S., and R. R. Reeves. 1982. Bottlenose dolphin *Tursiops truncatus* and other toothed cetaceans. In *Wild mammals of North America: Biology, management, and economics,* ed. J. A. Chapman and G. A. Feldhamer, 369–414. Baltimore: Johns Hopkins University Press.

———. 1983. Abundance of bottlenose dolphins in Corpus Christi Bay and coastal southern Texas. *Contrib. Mar. Sci.* 26:179–99.

Leatherwood, S., D. K. Caldwell, and H. E. Winn. 1976. Whales, dolphins, and porpoises of the western North Atlantic: A guide to their identification. *Nat. Ocean. Atm. Admin. Tech. Rep., Nat. Mar. Fish. Serv. Circ.* 396:1–176.

Leatherwood, S., J. R. Gilbert, and D. G. Chapman. 1978. An evaluation of some techniques for aerial censuses of bottlenosed dolphins. *J. Wildl. Manage.* 42:239–50.

Leatherwood, S., T. A. Jefferson, J. C. Norris, W. E. Stevens, L. J. Hansen, and K. D. Mullin. 1993. Occurrence and sounds of Fraser's dolphins (*Lagenodelphis hosei*) in the Gulf of Mexico. *Tex. J. Sci.* 43:349–54.

LeBoeuf, B. J., K. W. Kenyon, and B. Villa-Ramirez. 1986. The Caribbean monk seal is extinct. *Mar. Mammal Sci.* 2:70–72.

LeBoeuf, N., and D. Fertl. 1995. Sharing the sea. *Tide* (Magazine of the Gulf Coast Conservation Association, March/April) 6–8:50–51.

Lecke-Mitchell, K. M., and K. Mullin. 1992. Distribution and abundance of large floating plastic in the north–central Gulf of Mexico. *Mar. Pollut. Bull.* 24:598–601.

Lefebvre, L. W., T. J. O'Shea, G. B. Rathbun, and R. C. Best. 1989. Distribution, status, and biogeography of the West Indian manatee. In *Biogeography of the West Indies: Past, present, and future,* ed. C. A. Woods, 567–610. Gainesville, Fla.: Sandhill Crane Press.

Leonard, J., and R. D. Delumyea. 1994. Elemental carbon concentration in the air around Tampa Bay, Florida, 1990–1991. *Air & Waste* 44:804–806.

Lewis, J. 1996. Whales in the Gulf of Mexico. *La. Conserv.* March/April:4–7.

Lipka, D. A. 1970. The systematics and distribution of Enopleteuthidae and Cranchiidae (Cephalopoda: Oegopsida) from the Gulf of Mexico. M.S. thesis, Texas A&M University, College Station. 135 pp.

———. 1975. The systematics and zoogeography of cephalopods from the Gulf of Mexico. Ph.D. dissertation, Texas A&M University, College Station. 347 pp.

Lipscomb, T. P., S. Kennedy, D. Moffett, and B. K. Ford. 1994. Morbilliviral disease in an Atlantic bottlenose dolphin (*Tursiops truncatus*) from the Gulf of Mexico. *J. Wildl. Dis.* 30:572–76.

Lipscomb, T. P., S. Kennedy, D. Moffett, A. Krafft, B. A. Klaunberg, J. H. Lichy, G. T. Regan, G. A. J. Worthy, and J. K. Taubenberger. 1996. Morbilliviral epizootic in bottlenose dolphins of the Gulf of Mexico. *J. Vet. Diagn. Invest.* 8:283–90.

Lopez, A. M., D. B. McClellan, A. R. Bertolino, and M. D. Lange. 1979. The Japanese longline fishery in the Gulf of Mexico, 1978. *Mar. Fish. Rev.* 41(10):23–28.

Lowery, G. H. 1943. Check-list of the mammals of Louisiana and adjacent waters. *Occas. Pap. Mus. Zool.,* La. State Univ. 13:213–57.

———. 1974. *The mammals of Louisiana and its adjacent waters.* Baton Rouge: Louisiana State University Press. 588 pp.

LUMCON. 1995. *Effects of offshore oil and gas development: A current awareness bibliography.* Cocodrie: Louisiana Universities Marine Consortium. 17 pp.

Lutz, P. L., and J. A. Musick, eds. *The biology of sea turtles.* Boca Raton: CRC Press.

Lynn, S. K. 1995. Movements, site fidelity, and surfacing patterns of bottlenose dolphins on the central Texas coast. M.S. thesis, Texas A&M University, College Station.

Mahnken, T., and R. M. Gilmore. 1960. Suckerfish on porpoise. *J. Mammal.* 41:134.

Martin, N. 1988. You saw a what? *Tex. Shores* 21:26–29.

Mate, B. R., K. M. Stafford, and D. K. Ljungblad. 1994. A change in sperm whale (*Physeter macroephalus* [sic]) distribution correlated to seismic surveys in the Gulf of Mexico. *J. Acoust. Soc. Am.* 96(5, pt. 2):3268–69.

Mate, B. R., K. A. Rossbach, S. L. Nieukirk, R. S. Wells, A. B. Irvine, M. D. Scott, and A. J. Read. 1995. Satellite-monitored movements and dive behavior of a bottlenose dolphin (*Tursiops truncatus*) in Tampa Bay, Florida. *Mar. Mammal Sci.* 1:452–63.

Maze, K. S. 1997. Bottlenose dolphins of San Luis Pass, Texas: Occurrence patterns, site fidelity, and habitat use. M.S. thesis, Texas A&M University, College Station. 79 pp.

McHugh, M. B. 1989. Population numbers and feeding behavior of the Atlantic bottlenose dolphin (*Tursiops truncatus*) near Aransas Pass, Texas. M.S. thesis, University of Texas, Austin.

Mead, J. G. 1973. Marine mammal strandings and sightings investigated. *Underwat. Nat.* 8:12–13.

———. 1977. Records of sei and Bryde's whales from the Atlantic Coast of the United States, the Gulf of Mexico, and the Caribbean. *Rep. Int. Whaling Comm.* Special Issue 1:113–16.

———. 1979. An analysis of cetacean strandings along the eastern coast of the United States. In *Biology of marine mammals: Insights through strandings,* ed. J. R. Geraci and D. J. St. Aubin, 54–68. Washington, D.C.: National Technical Information Service.

———. 1986. Twentieth-Century records of right whales (*Eubalaena glacialis*) in the northwestern Atlantic Ocean. *Rep. Int. Whaling Comm.* Special Issue 10:109–19.

Mead, J. G., and C. W. Potter. 1990. Natural history of bottlenose dolphins along the central Atlantic Coast of the United States. In *The bottlenose dolphin,* ed. S. Leatherwood and R. R. Reeves, 165–95. San Diego: Academic Press.

Mead, J. G., D. K. Odell, R. S. Wells, and M. D. Scott. 1980. Observations on a mass stranding of spinner dolphins, *Stenella longirostris,* from the west coast of Florida. *Fish. Bull.* 78:353–60.

Messinger, C. 1989. A fight for life. *Whalewatcher* 23:15–17.

Miller, G. S. 1920. American records of whales of the genus *Pseudorca. Proc. U.S. Nat. Mus.* 57:205–207.

———. 1924. A pollack whale from Florida presented to the National Museum by the Miami Aquarium Association. *Proc. U.S. Nat. Mus.* 66:1–15.

———. 1924. List of North American recent mammals 1923. *Proc. U.S. Nat. Mus.* 128:1–673.

———. 1928. The pollack whale in the Gulf of Campeche. *Proc. Biol. Soc. Wash.* 41:171.

Miller, G. S., and R. Kellogg. 1955. List of North American recent mammals. *Bull. U.S. Nat. Mus.* 205:1–954.

Miller, G. W. 1992. An investigation of dolphin (*Tursiops truncatus*) deaths in east Matagorda Bay, Texas, January 1990. *Fish. Bull.* 90:791–97.

Mills, L. R., and K. R. Rademacher. 1996. Atlantic spotted dolphins (*Stenella frontalis*) in the Gulf of Mexico. *Gulf Mar. Sci.* 1996:114–20.

Mitchell, E. D. 1991. Winter records of the minke whale (*Balaenoptera acutorostrata* Lacepede 1804) in the southern North Atlantic. *Rep. Int. Whaling Comm.* 41:455–57.

Moore, J. C. 1953. Distribution of marine mammals in Florida waters. *Am. Midl. Nat.* 49:117–58.

———. 1955. Bottle-nosed dolphins support remains of young. *J. Mammal.* 36:466–67.

———. 1958. A beaked whale from the Bahama Islands and comments on the distribution of *Mesoplodon densirostris. Am. Mus. Novitiates* 1897:1–12.

———. 1960. New records of the Gulf-stream beaked whale, *Mesoplodon gervaisi,* and some taxonomic considerations. *Am. Mus. Novitiates* 1993:1–35.

———. 1963. The goose-beaked whale: Where in the world? *Chi. Nat. Hist. Mus. Bull.* 34:2–3, 8.

Moore, J. C., and E. Clark. 1963. Discovery of right whales in the Gulf of Mexico. *Science* 141:269.

Moore, J. C., and S. Palmer. 1955. More piked whales from the southern North Atlantic. *J. Mammal.* 36:429–33.

Moore, J. C., and F. G. Wood. 1957. Differences between the beaked whales *Mesoplodon mirus* and *Mesoplodon gervaisi. Am. Mus. Novitiates* 1831:5–25.

Mullin, K. D. 1988. Comparative seasonal abundance of bottlenose dolphins (*Tursiops truncatus*) in three habitats of the north–central Gulf of Mexico. Ph.D. dissertation, Mississippi State University.

Mullin, K. D., and L. J. Hansen. In press. Marine mammals in the northern Gulf of Mexico. In *Gulf of Mexico: A large marine ecosystem,* ed. T. D. McIlwain and H. E. Kumph. Symposium Proceedings.

Mullin, K. D., L. V. Higgins, T. A. Jefferson, and L. J. Hansen. 1994. Sightings of the Clymene dolphin (*Stenella clymene*) in the Gulf of Mexico. *Mar. Mammal Sci.* 10:464–70.

Mullin, K. D., T. A. Jefferson, L. J. Hansen, and W. Hoggard. 1994. First sighting of melon-headed whales (*Peponocephala electra*) in the Gulf of Mexico. *Mar. Mammal Sci.* 10:342–48.

Mullin, K. D., R. R. Lohoefener, W. Hoggard, C. L. Roden, and C. M. Rogers. 1990. Abundance of bottlenose dolphins, *Tursiops truncatus,* in the coastal Gulf of Mexico. *Northeast Gulf Sci.* 11:113–22.

Mullin, K. D., W. Hoggard, C. Roden, R. Lohoefener, C. Rogers, and B. Taggart. 1991. *Cetacaens on the upper Continental Slope in the north–central Gulf of Mexico.* OCS Study/MMS 91-0027. New Orleans: U.S. Department of the Interior, Minerals Management Service, Gulf of Mexico OCS Regional Office.

———. 1994. Cetaceans on the upper Continental Slope in the north–central Gulf of Mexico. *Fish. Bull.* 92:773–86.

Mullin, K. D., R. R. Lohoefener, W. Hoggard, C. M. Rogers, C. L. Roden, and B. Taggart. 1991. Whales and dolphins offshore of Alabama. *J. Ala. Acad. Sci.* 62:48–58.

National Marine Fisheries Service. 1994. *Report to Congress on results of feeding wild dolphins: 1989–1994.* Silver Springs, Md.

Navarro-L., D. 1988. A stranding of *Globicephala macrorhynchus* (Cetacea: Delphinidae) in Yucatan, Mexico. *Southwest. Nat.* 33:247–48.

Negus, N. C., and R. K. Chipman. 1956. A record of the piked whale, *Balaenoptera acutorostrata,* off the Louisiana coast. *Proc. La. Acad. Sci.* 19: 41–42.

Nelson, D. M., ed. 1992. *Distribution and abundance of fishes and invertebrates in Gulf of Mexico estuaries.* Vol. 1: *Data summaries.* ELM Report No. 10. Rockville, Md.: NOAA/NOS Strategic Environmental Assessments Division.

Newman, H. H. 1910. A large sperm whale captured in Texas waters. *Science* 31:631–32.

Norris, J., W. Evans, T. Sparks, and R. Benson. 1995. The use of passive towed arrays for surveying marine mammals. *J. Acoust. Soc. Am.* 97(5):3353.

Odell, D. K. 1991. A review of the southeastern United States marine mammal stranding network: 1978–1987. *Nat. Ocean. Atm. Admin. Tech. Rep., Nat. Mar. Fish. Serv.* 98:19–23.

Odell, D. K., D. B. Sniff, and G. H. Waring, eds. 1975. *Tursiops truncatus Assessment Workshop, final report.* Marine Mammal Commission Contract No. MM5AC021. Silver Springs, Md.

Odell, D. K., M. T. Walsh, and E. D. Asper. 1989. Cetacean mass strandings: Healthy vs. sick animals. *Whalewatcher* 23:9–10.

Odell, D. K., E. D. Asper, J. Baucom, and L. H. Cornell. 1980. A recurrent mass stranding of the false killer whale, *Pseudorca crassidens,* in Florida. *Fish. Bull.* 78:171–76.

O'Shea, T. J., B. L. Homer, E. C. Greiner, and A. W. Layton. 1991. *Nasitrema* sp.-associated encephalitis in a striped dolphin (*Stenella coeruleoalba*) stranded in the Gulf of Mexico. *J. Wildl. Dis.* 27:706–709.

O'Sullivan, S., and K. D. Mullin. 1997. Killer whales (*Orcinus orca*) in the northern Gulf of Mexico. *Mar. Mammal Sci.* 13:141–47.

Paul, J. R. 1968. Risso's dolphin, *Grampus griseus,* in the Gulf of Mexico. *J. Mammal.* 49:746–47.

Perrin, W. F., E. D. Mitchell, J. G. Mead, D. K. Caldwell, and P. J. H. van Bree. 1981. *Stenella clymene,* a rediscovered tropical dolphin of the Atlantic. *J. Mammal.* 62:583–98.

Perrin, W. F., E. D. Mitchell, J. G. Mead, D. K. Caldwell, M. C. Caldwell, P. J. H. van Bree, and W. H. Dawbin. 1987. Revision of the spotted dolphins, *Stenella* spp. *Mar. Mammal Sci.* 3:99–170.

Perryman, W. L., D. K. Au, S. Leatherwood, and T. A. Jefferson. 1994. Melon-headed whale *Peponocephala electra* Grey, 1846. In *Handbook of marine mammals.* Vol. 5: *The first book of dolphins,* ed. S. H. Ridgway and R. Harrison, 363–86. London: Academic Press.

Peterson, J. C., and W. Hoggard. 1996. First sperm whale (*Physeter macrocephalus*) record in Mississippi. *Gulf Res. Rep.* 9(3):215–17.

Porter, J. W. 1977. *Pseudorca* stranding. *Oceans* 10:8–16.

Powell, J. A., and G. B. Rathbun. 1984. Distribution and abundance of manatees along the northern coast of the Gulf of Mexico. *Northeast Gulf Sci.* 7:1–28.

Rankin, J. J. 1953. First record of the rare beaked whale, *Mesoplodon europaeus,* Gervais, from the West Indies. *Nature* 172:873.

———. 1955. A rare whale in tropical seas. *Everglades Nat. Hist.* 3:24–31.

———. 1956. The structure of the skull of the beaked whale, *Mesoplodon gervaisi. J. Morphol.* 99:329–57.

Raun, G. G. 1964. West Indian seal remains from two historic sites in coastal south Texas. *Bull. Tex. Arch. Soc.* 35:189–92.

Raun, G. G., H. D. Hoese, and F. Moseley. 1970. Pygmy sperm whales, genus *Kogia,* on the Texas coast. *Tex. J. Sci.* 21:269–74.

Rawson, A. J., G. W. Patton, H. F. Anderson, and T. Beecher. 1991. Anthracosis in the Atlantic bottlenose dolphin (*Tursiops truncatus*). *Mar. Mammal Sci.* 7:413–16.

Rawson, A. J., J. P. Bradley, A. Teetsov, S. B. Rice, E. M. Haller, and G. W. Patton. 1995. A role for airborne particulates in high mercury levels of some cetaceans. *Ecotoxicol. Environ. Safe.* 30:309–14.

Read, A. J., R. S. Wells, A. A. Hohn, and M. D. Scott. 1993. Patterns of growth in wild bottlenose dolphins, *Tursiops truncatus. J. Zool.,* London 231:107–23.

Reeves, R. R., J. G. Mead, and S. Katona. 1978. The right whale, *Eubalaena glacialis,* in the western North Atlantic. *Rep. Int. Whaling Comm.* 28:303–12.

Regan, G. T. 1990. Geography, seasonality, size, and sex of bottlenose dolphins found dead in Alabama since 1987. *J. Ala. Acad. Sci.* 61:144.

———. 1991. A sharp rise in the number of bottlenose dolphins found dead in Alabama. *J. Ala. Acad. Sci.* 62:82.

Reynolds, J. E., III. 1985. *Evaluation of the nature and magnitude of interactions between bottlenose dolphins, Tursiops truncatus, and fisheries and other human activities in the coastal areas of the southeastern United States.* PB86-162-203. Washington, D.C.: National Technical Information Service.

Ribic, C. A., R. Davis, N. Hess, and D. Peake. 1997. Distribution of seabirds in the northern Gulf of Mexico in relation to mesoscale features: Initial observations. *ICES J. Mar. Sci.* 54:545–51.

Rice, D. W. 1965. Bryde's whale in the Gulf of Mexico. *Norsk Hvalfangst-Tidende* 54:114–15.

Richard, K. R., and M. A. Barbeau. 1994. Observations of spotted dolphins feeding nocturnally on flying fish. *Mar. Mammal Sci.* 10:473–77.

Ridgway, S. H., and K. W. Benirschke, eds. 1977. *Breeding dolphins: Present status, suggestions for the future.* PB-273-673. Springfield, Va.: United States Department of Commerce, National Technical Information Service.

Ripple, J. 1996. *Sea turtles.* Stillwater, Minn.: Voyageur Press. 84 pp.

Russell, S. J. 1993. Shark bycatch in the northern Gulf of Mexico tuna longline fishery, 1988–91, with observations on the nearshore directed shark fishery. *Nat. Ocean. Atm. Admin. Tech. Rep., Nat. Mar. Fish. Serv.* 115:19–29.

Salata, G. G. 1993. Analysis of Gulf of Mexico marine mammals for organochlorine pesticides and PCBs. M.S. thesis, Texas A&M University, College Station. 134 pp.

Salata, G. G., T. L. Wade, J. L. Sericano, J. W. Davis, and J. M. Brooks. 1995.

Analysis of Gulf of Mexico bottlenose dolphins for organochlorine pesticides and PCBs. *Environ. Pollut.* 88:167–75.

Salinas Z., M., A. Aguayo L., C. Alvarez F., I. Fuentes A., and B. Morales V. 1984. *Observaciones de cetáceos a bordo del B/O Justo Sierra durante las campaña Chapo I (Agosto de 1983) y Chapo II (Octubre–Noviembre de 1983).* Mexico City: Universidad Nacional Autónoma de México Scientific Report.

Sayigh, L. S., P. L. Tyack, and R. S. Wells. 1993. Recording underwater sounds of free-ranging dolphins while underway in a small boat. *Mar. Mammal Sci.* 9:209–12.

Sayigh, L. S., P. L. Tyack, R. S. Wellsand, and M. D. Scott. 1990. Signature whistles of free-ranging bottlenose dolphins *Tursiops truncatus*: Stability and mother–offspring comparisons. *Behav. Ecol.* 26:247–60.

Sayigh, L. S., P. L. Tyack, R. S. Wellsand, M. D. Scott., and A. B. Irvine. 1995. Sex difference in signature whistle production of free-ranging bottlenose dolphins, *Tursiops truncatus. Behav. Ecol. Sociobiol.* 36:171–77.

Schiro, A., and D. Fertl. 1995. Mermaids sighted in Galveston Bay. *Soundings* 7:4–5.

Schiro, A., L. P. May, and D. C. Fertl. 1996. *Manatee occurrence in the northwestern Gulf of Mexico* (Abstract). November 8–11. San Pedro, Calif.: American Cetacean Society Conference.

Schiro, A., B Würsig, D. Weller, J. Norris, and M. Armstrong. 1995. *Preliminary findings on sightings and movements of individually identified sperm whales (Physeter macrocephalus) in the north central Gulf of Mexico* (Abstract). Meeting, March 3–4. Waco: Texas Academy of Science.

Schmidly, D. J. 1981. *Marine mammals of the southeastern United States coast and the Gulf of Mexico.* FWS/OBS-80/41. Washington, D.C.: U.S. Fish and Wildlife Service, Office of Biological Services. 163 pp.

Schmidly, D. J., and B. A. Melcher. 1974. Annotated checklist and key to the cetaceans of Texas waters. *Southwest. Nat.* 18:453–64.

———. 1974. Whales, porpoises, and dolphins of Texas. *Tex. Parks Wildl.* 32(1):12–15, 32(2):12–14, 32(3):18–22.

Schmidly, D. J., and S. H. Shane. 1978. *A biological assessment of the cetacean fauna of the Texas coast.* PB-281-763. Springfield, Va.: National Technical Information Service. 38 pp.

Schmidly, D. J., M. H. Beleau, and H. Hildebrand. 1972. First record of Cuvier's dolphin from the Gulf of Mexico, with comments on the taxonomic status of *Stenella frontalis. J. Mammal.* 53:625–28.

Schmidly, D. J., C. O. Martin, and G. F. Collins. 1972. First occurrence of a black right whale (*Balaena glacialis*) along the Texas coast. *Southwest. Nat.* 17:214–15.

Scott, G. P. 1990. Management-oriented research on bottlenose dolphins by the Southeast Fisheries Center. In *The bottlenose dolphin,* ed. S. Leatherwood and R. R. Reeves, 623–39. San Diego: Academic Press.

Scott, G. P., D. M. Burn, L. J. Hansen, and R. E. Owen. 1989. *Estimates of bottlenose dolphin abundance in the Gulf of Mexico from regional aerial surveys.* Miami: National Marine Fisheries Service, Southeast Fisheries Science Center.

Scott, M. D., R. S. Wells, and A. B. Irvine. 1990. A long-term study of bottlenose dolphins on the west coast of Florida. In *The bottlenose dolphin,* ed. S. Leatherwood and R. R. Reeves, 235–44. San Diego: Academic Press.

Shane, S. H. 1977. The population biology of the Atlantic bottlenose dolphin, *Tursiops truncatus,* in the Aransas Pass area of Texas. M.S. thesis, Texas A&M University, College Station. 257 pp.

———. 1978. Suckerfish attached to a bottlenose dolphin in Texas. *J. Mammal.* 59:439–40.

———. 1980. Occurrence, movements, and distribution of the bottlenose dolphin, *Tursiops truncatus,* in southern Texas. *Fish. Bull.* 78:593–601.

———. 1987. The behavioral ecology of the bottlenose dolphin. Ph.D. dissertation, University of California at Santa Cruz. 154 pp.

———. 1988. *The bottlenose dolphin in the wild.* San Carlos, Calif.: Hatcher Trade Press.

———. 1990. Behavior and ecology of the bottlenose dolphin at Sanibel Island, Florida. In *The bottlenose dolphin,* ed. S. Leatherwood and R. R. Reeves, 245–65. San Diego: Academic Press.

———. 1990. Comparison of bottlenose dolphin behavior in Texas and Florida, with a critique of methods for studying dolphin behavior. In *The bottlenose dolphin,* ed. S. Leatherwood and R. R. Reeves, 541–58. San Diego: Academic Press.

Shane, S. H., and D. J. Schmidly. 1976. Bryde's whale (*Balaenoptera edeni*) from the Louisiana coast. *Southwest. Nat.* 21:409–10.

———. 1978. *The population biology of the Atlantic bottlenose dolphin, Tursiops truncatus, in the Aransas Pass area of Texas.* PB-283-393. Washington, D.C.: National Technical Information Service.

Shane, S. H., R. S. Wells, and B. Würsig. 1986. Ecology, behavior, and social organization of the bottlenose dolphin: A review. *Mar. Mammal Sci.* 2:34–63.

Smultea, M. A., and B. Würsig. 1995. Behavioral reactions of bottlenose dolphins to the Mega Borg II oil spill, Gulf of Mexico. *Aquat. Mammals* 21:171–83.

Sparks, T. D. 1997. Distributions of sperm whales along the Continental Slope in the northwestern and central Gulf of Mexico as determined from an acoustic survey. M.S. thesis, Texas A&M University, College Station. 67 pp.

Springer, S. 1957. Some observations on the behavior of schools of fishes in the Gulf of Mexico and adjacent waters. *Ecology* 38:166–71.

Stearns, S. 1887. Fisheries of the Gulf of Mexico. In *The fisheries and fishery industries of the United States.* Sec. II. *A geographical review of the fishery industries and fishery communities for the year 1880,* pt. XV, ed. G. B. Goode, 533–87. Washington, D.C.: U.S. Government Printing Office.

Stephens, W. M. 1965. Sperm whale stranded. *Sea Frontiers* 11:362–67.

Suzik, H. A. 1997. Unraveling the manatee mystery. *J. Am. Vet. Med. Assoc.* 210:740.

Tarpley, R. J. 1987. Texas marine mammal stranding network. *Southwest. Vet.* 38:51–58.

Tarpley, R. J., and S. Marwitz. 1986. The mystery of stranded marine mammals. *Tex. Parks Wildl.* 44:32–38.

———. 1993. Plastic ingestion by cetaceans along the Texas coast: Two case reports. *Aquat. Mammals* 19:93–98.

Tolley, K. A., A. J. Read, R. S. Wells, K. W. Urian, M. D. Scott, A. B. Irvine, and A. A. Hohn. 1995. Sexual dimorphism in a community of bottlenose dolphins (*Tursiops truncatus*) from Sarasota, Florida. *J. Mammal.* 76:1190–98.

Townsend, C. H. 1906. Capture of the West Indian seal (*Monachus tropicalis*) at Key West, Florida. *Science* 23:583.

———. 1935. The distribution of certain whales as shown by logbook records of American whaleships. *Zoologica* 19:3-50.

True, F. W. 1885. A rare dolphin. *Science* 6:44.

———. 1885. On a spotted dolphin apparently identical with the *Prodelphinus doris* of Gray. *Annu. Rep. Smithsonian Inst.* 1884:317–24.

Tucker & Associates, Inc. 1990. *Sea turtles and marine mammals of the Gulf of Mexico.* Proceedings of a workshop held in New Orleans, August 1–3, 1989. OCS Study MMS 90-0009. New Orleans: U.S. Department of the Interior, Minerals Management Service, Gulf of Mexico OCS Region.

Uchupi, E. 1975. Physiography of the Gulf of Mexico and Caribbean Sea. In *The ocean basins and margins.* Vol. 3: *The Gulf of Mexico and the Caribbean,* ed. A. E. M. Nairn and F. G. Stehli, 1–64. New York: Plenum Press.

United States Fish and Wildlife Service. 1995. *Florida manatee recovery plan,* 2d revision. Atlanta: U.S. Fish and Wildlife Service. 160 pp.

Urbán-R. J., and A. Aguayo-L. 1983. *Observaciones de mamíferos marinos a bordo del B/O Justo Sierra durante la campaña oceanográfica Yucatan* I. *Mayo de 1983.* Mexico City: Universidad Nacional Autonoma de México.

Urian, K. W., D. A. Duffield, A. J. Read, R. S. Wells, and E. D. Shell. 1996. Seasonality of reproduction in bottlenose dolphins, *Tursiops truncatus. J. Mammal.* 77:394–403.

Varona, L. S. 1964. Un craneo de Ziphius cavirostris del sur Isla de Piños. *Poeyana* 4A:1–3.

———. 1970. Morphología externa y caracteres craneales de un macho adulto de *Mesoplodon europaeus* (Cetacea: Ziphiidae). *Poeyana* 69A:1–17.

Voss, G. L. 1956. A review of the cephalopods of the Gulf of Mexico. *Bull. Mar. Sci. Gulf Caribb.* 6(2):85–178.

———. 1971. The cephalopod resources of the Caribbean Sea and adjacent regions. *FAO Fish. Rep.* 71(2):307–23.

Vokuvich, F. M., and B. W. Crissman. 1986. *Aspects of warm rings in the Gulf of Mexico.* J. Geophys. Res. 91:2645–60.

Waldo, E. 1957. Whales in the Gulf of Mexico. *La. Conserv.* 9:14–15.

Wang Ding, B. Würsig, and W. Evans. 1995. Comparisons of whistles among seven odontocete species. In *Sensory systems of aquatic mammals,* ed. R. A. Kastelein, J. A. Thomas, and P. E. Nachtigall, 299–324. Woerden: De Spil Publishers.

———. 1995. Whistles of bottlenose dolphins: Comparisons among populations. *Aquat. Mammals* 21:65–77.

Weber, M., R. T. Townsend, and R. Bierce. 1992. *Environmental quality in the*

*Gulf of Mexico: A citizen's guide,* 2d edition. Washington, D.C.: Center for Marine Conservation.

Weigle, B. L. 1987. Abundance, distribution, and movements of bottlenose dolphins (*Tursiops truncatus*) in lower Tampa Bay, Florida. M.S. thesis, University of South Florida, St. Petersburg.

———. 1990. Abundance, distribution, and movements of bottlenose dolphins (*Tursiops truncatus*) in lower Tampa Bay, Florida. *Rep. Int. Whaling Comm.* Special Issue 14:195–202.

Weller, D. W., A. J. Schiro, V. G. Cockcroft, and W. Ding. 1996. First account of a humpback whale (*Megaptera novaeangliae*) in Texas waters, with a re-evaluation of historical records from the Gulf of Mexico. *Mar. Mammal Sci.* 12:133–37.

Weller, D. W., B. Würsig, H. Whitehead, J. Norris, S. Lynn, R. Davis, N. Clauss, and P. Brown. 1996. Observations of an interaction between sperm whales and short-finned pilot whales in the Gulf of Mexico. *Mar. Mammal Sci.* 12:588–93.

Wells, R. S. 1978. Home range characteristics and group composition of Atlantic bottlenosed dolphins, *Tursiops truncatus,* on the west coast of Florida. M.S. thesis, University of Florida, Gainesville.

———. 1986. Structural aspects of dolphin societies. Ph.D. dissertation, University of California at Santa Cruz.

———. 1991. Bringing up baby. *Nat. Hist.* 100(8):56–62.

———. 1991. The role of long-term study in understanding the social structure of a bottlenose dolphin community. In *Dolphin societies: Discoveries and puzzles,* ed. K. Pryor and K. S. Norris, 199–225. Berkeley: University of California Press.

———. 1992. *The marine mammals of Sarasota Bay.* 1992 Framework for action. Sarasota, Fla.: Sarasota Bay National Estuary Program.

Wells, R. S., and M. D. Scott. 1990. Estimating bottlenose dolphin population parameters from individual identification and capture–release techniques. *Rep. Int. Whaling Comm.* Special Issue 14:407–15.

———. 1997. Seasonal incidence of boat strikes on bottlenose dolphins near Sarasota, Florida. *Mar. Mammal Sci.* 13:475–80.

Wells, R. S., A. B. Irvine, and M. D. Scott. 1980. The social ecology of inshore odontocetes. In *Cetacean behavior: Mechanisms and functions,* edited by L. M. Herman, 263–317. New York: John Wiley & Sons.

Wells, R. S., M. D. Scott, and A. B. Irvine. 1987. The social structure of free-ranging bottlenose dolphins. In *Current mammalogy,* vol. 1, ed. H. H. Genoways, 247–305. New York: Plenum Press.

Wells, R. S., M. K. Bassos, K. W. Urian, W. J. Carr, and M. D. Scott. 1996. Low-level monitoring of bottlenose dolphins, *Tursiops truncatus,* in Charlotte Harbor, Florida 1990–1994. *Nat. Ocean. Atm. Admin. Tech. Memo., Nat. Mar. Fish. Serv., Southeast Fish. Sci. Ctr.* 384:1–36.

Wells, R. S., K. W. Urian, A. J. Read, M. K. Bassos, W. J. Carr, and M. D. Scott. 1996. Low-level monitoring of bottlenose dolphins, *Tursiops truncatus,* in Tampa Bay, Florida 1988–1993. *Nat. Ocean. Atm. Admin. Tech. Memo., Nat. Mar. Fish. Serv., Southeast Fish. Sci. Ctr.* 385:1–25.

Wood, F. G. 1953. Underwater sound production and concurrent behavior of captive porpoise, *Tursiops truncatus,* and *Stenella plagiodon. Bull. Mar. Sci. Gulf Caribb.* 3:120–33.

Wood, F. G., Jr., D. K. Caldwell, and M. C. Caldwell. 1970. Behavioral interactions between porpoises and sharks. *Invest. Cetacea* 2:264–77.

Würsig, B. 1991. Cooperative foraging strategies: An essay on dolphins and us. *Whalewatcher* 25:3–6.

Würsig, B., and T. Henningsen. 1991. The bottlenose dolphins of Galveston shores. *Galveston Bay Soundings* 3:13.

Würsig, B., and S. K. Lynn. 1996. Movements, site fidelity, and respiration patterns of bottlenose dolphins on the central Texas coast. *Nat. Ocean. Atm. Admin. Tech. Memo., Nat. Mar. Fish. Serv., Southeast Fish. Sci. Ctr.* 383:1–111.

Würsig, B., S. K. Lynn, T. A. Jefferson, and K. D. Mullin. 1998. Behavior of cetaceans in the northern Gulf of Mexico relative to survey ships and aircraft. *Aquat. Mammals* 24:41–50.

Zam, S. G., D. K. Caldwell, and M. C. Caldwell. 1971. Some endoparasites from small odontocete cetaceans collected in Florida and Georgia. *Cetology* 2:1–11.

Zarate Becerra, E. 1993. Distribución del manati (Trichechus manatus) en la porción sur de Quintana Roo, México. *Revista de Investigación Científica* 1:1–12.

Zavala-González, A., J. Urbán-Ramírez, and C. Esquivel-Macías. 1994. A note on artisanal fisheries interactions with small cetaceans in Mexico. *Rep. Int. Whaling Comm.* Special Issue 15:235–40.

Zimmerman, R. J. 1994. *A preliminary report on mortalities of marine animals during the spring of 1994.* Galveston: National Marine Fisheries Service.

# Index

Note: Pages with illustrations are indicated by italics and maps are indicated by bold. Items in the color photo gallery are designated with "gallery" followed by the photo number, while the artist renderings are designated with "plate" followed by the plate number.